T0258212

SPORT

JAY SCHULKIN

SPORT

—

A Biological,
Philosophical, and
Cultural Perspective

COLUMBIA UNIVERSITY PRESS
NEW YORK

Columbia University Press
Publishers Since 1893
New York   Chichester, West Sussex
cup.columbia.edu
Copyright © 2016 Columbia University Press

Library of Congress Cataloging-in-Publication Data
Names: Schulkin, Jay.
Title: Sport : a biological, philosophical, and cultural perspective / Jay Schulkin.
Description: New York : Columbia University Press, [2016] |
Includes bibliographical references and index.
Identifiers: LCCN 2015048090 (print) | LCCN 2016025084 (ebook) |
ISBN 9780231176767 (cloth : alk. paper) | ISBN 9780231541978 (e-book) |
ISBN 9780231541978
Subjects: LCSH: Sports—Physiological aspects. | Sports—Social aspects.
Classification: LCC RC1235 .S394 2016 (print) | LCC RC1235 (ebook) |
DDC 613.7—dc23
LC record available at https://lccn.loc.gov/2015048090

Columbia University Press books are printed on permanent
and durable acid-free paper.
Printed in the United States of America

c 10 9 8 7 6 5 4 3 2 1

Cover design: Elliot Strunk/Fifth Letter
Cover image: ©iStockphoto

This book is dedicated to the Siegelman family,
Russ, Beth, Max, and Jacob, for the friendship
and generosity they extended to my family.

And to my colleague, friend, and
fellow New York fan Alex Martin.

I also want to thank my children's coaches, who have helped
them develop a love for sports.

CONTENTS

INTRODUCTION

A thing of brilliance, brawn, and beauty, sport is as natural an occurrence as breathing and language. For some of us, one sport or another—seasonal or continuous, and whether or not we formally classify it as sport—is what we look to as a source of relief, filling up the space of life, a pastime that can transcend the stuff that divides us.

I am not an athlete, but I am physical and coordinated. When I lived in New York City as a young man and through my thirties, I walked everywhere. While I did not engage in formal athletic training, I have spent many years in the company of athletes, one in particular: my wife of almost twenty-seven years.

The first thing my wife, a swimmer, noticed when she came to my NYC apartment before our marriage was the swimming and sports facility in my building. I knew it was there; I just had never stepped inside. I didn't have a reason to. But, of course, she did. Her world was the discipline and the joy and the pain of formal sport and training.

While I may not be an athlete like my wife, my world clearly includes being physical, and I am certainly driven. Physicality, stamina, and drive are key features for any sporting activity. Walking, thinking, and looking were the constants in my world, and walking—and walking quickly—is being physical. We might even regard it as the evolutionary origin of sport, when we climbed down out of the trees, ventured out onto the open plains, and eventually developed into the only animal that practices sustained and purposeful distance running. In fact, walking can be competitive and is an Olympic sport. Now, in my opinion (and I mean no disrespect), competitive walking is not pretty to look at; the hip business

FIGURE 0.1

Butterfly stroke.

doesn't do it for me. But the simple act of physical movement does, and walking was my way to accomplish that.

Though I didn't physically engage in sports, mentally it was always one of my passions. As a young man I followed sport news and scores: the Yankees, the Knicks, the Giants. At that time, I was part of a neuroscientific laboratory at New York University in Washington Square Park. I would often come out to watch guys play pickup ball. Many times, these were men who had almost made the big leagues, and every once in a while their professional buddies would show up and play with them. They were wonderful to watch, and their trash talk and bravado added to the air of excitement. It was a bit surreal watching such games happen in Greenwich Village, NY, but when these games did take place, there was some great sport on the floor.

When one of my teams won, I could not read enough about them, and when they lost, I wanted to avoid it. Losing is hard even when you're not playing. Working hard and displaying character in losing are qualities of the ethics of sport. But keeping the mind sharp is equally important: Being able to keep track of statistics is crucial in scorekeeping, picking the roster, and following games such as baseball. Humility is also a necessary trait for the player: Batting .300 is a good average, but it actually means that the majority of the time the player is not hitting the ball but is striking out, popping out, or grounding out. There is also a lot of courage involved when someone is throwing a ball at you. Sport is as much a mental and moral exercise as a physical one.

Baseball, while rich for me, was boring for my wife until our son started playing. She got involved, learned the game, and became a fan. My son's wonderful coach, who was from Boston, was a major Boston Red Sox fan. He was a catcher, as was my son, and he taught our whole family much about the game and the whole mentality of sport. Watching our children engage in athletic activities has given us lifelong friends and engendered many fond memories. Sport is as much social as it is physical.

Sport is large and primordial. It can be used for good or bad. Hitler glorified his vision of the Nazi elite athlete at the 1936 Olympics (though he was stymied by the African American track star Jesse Owens). Nelson Mandela (a hero to many of us), on the other hand, who stressed that "sport has the power to change the world," used the 1995 Rugby World Cup to unite a racially divided South Africa. Sport goes far beyond winning or losing a specific event. It is a summary of the human condition.

So what does sport do for us? Sport plays a role in our childhood development via the biology associated with play and pedagogy. Pedagogy and expanded play are linked to social contact, and social contact is our foothold in the world around us. Sport emerged from capabilities that are embedded in culture, and participating in sport (in its many forms) is a part of becoming a member of a culture. As Margaret Mahler (2000) intimated, there is biological birth, psychological birth, and social birth, and sport is tied to a sense of social birth; sport enhances the biological capabilities that we bring with us into the world. The cultural is continuous with the biological, and neither reduces the other.

Sport traverses every part of the child, varying, of course, with the culture that child is a part of. The development of form and function is essential to sport, and it is gained by practice and play. These are important components in the normal development of one's skills and personal sense of competence and excellence. Finally, there is a sense of achievement and discovery associated with participation in sports.

Sport evolved, in part, to facilitate our socialization and our sense of belonging to a particular group. It is also an ancient practice. Several different cultures (for example, China and northern Europe) can independently trace skiing as a practice for more than two thousand years. Sport is culturally ancient and is all-pervasive in the modern era.

What can you expect to gain from this book? First, an appreciation of sport as a part of life that is as important and specific to us as language, standing up straight, singing, or agriculture. Second, an awareness that,

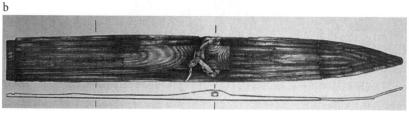

FIGURE 0.2

(a) One of the two figures of skiers carved in stone near Rödöy, Norway (ca. 2000 BC). This is the oldest known reference to skiing. (b) The oldest known ski is shown as it was found; the remaining band that passed around the heel is only partially visible.

Source: National Board of Antiquities, Helsinki, Finland.

Adapted from Formenti et al. (2005).

as in most things in life, diverse biological conditions underlie different forms of sport. Third, the recognition that though it functions as a kind of universal language, the expression of sport varies among cultures. Fourth, the realization that to understand the biology and neuroscience of sport is to understand something fundamental about us as a species.

In this book, mind, body, and culture harmonize on a continuous theme. We will learn that within sport exist core features of biology and neural systems tied to adaptation and action. The same physical and intellectual components that underlie diverse adaptations outside of sport (language; spatial and temporal capabilities; inference; memory; agency and direction; causation; detection of intention of others; mathematical calculation; endurance; etc.) underlie sport as well.

Why this book? No other work makes clear the biology that underlies sport. From the evolution of our species and our brain's functions, to the diverse information molecules (dopamine, endorphins, oxytocin) our body employs, to the expansion and flexibility of the shoulder muscle (for throwing), to our expanded Achilles tendon (for standing erect and running), I suggest that sport is a really good example of the continuity of biological and cultural evolutionary trends.

Thus this book is written from a biological perspective, and evolutionary considerations figure largely in it. I have always been influenced by John Dewey and the classical pragmatists, which are reflected throughout this book. Key events that the book addresses include our sense of the body, our bipedal stance, our big brains, our distribution across the earth, and our rich endowment with social capabilities. While writing this book, I often thought about two amazing teachers, Paul Weiss and Sophia Delza. Paul Weiss was a philosopher and someone I studied with, and on many things disagreed with, long ago. He wrote a book titled *Sport: A Philosophical Inquiry* in which he recognized and elegantly wrote about the training, sacrifice, dedication, discipline, endurance, and motivation that goes into the pursuit of athletic excellence. Sophia Delza, a dancer and martial-arts expert, was all about perfection, form, and discipline; a mind very much in the body. She was a wonderful teacher.

This is a short book, and it suggests more than I have room to demonstrate. I apologize in advance for any individual or topic not included in this book. This is a small window into the rich world of sports. But I know of no subject that more captures the human drama in all its features—the good, the bad, and the ugly—than sport. Finally, with love, I would like to thank my three athletes: April Oliver and Danielle and Nick Schulkin.

SPORT

1

THE CONCEPT OF SPORT

ristotle hypothesized that the "unmoved mover" was the highest state of being. Many other early philosophers—Buddha, for instance—believed something similar. This may seem counterintuitive to us. Movement is, after all, essential to life. In everyday speech we associate movement with both beauty and capability: a person runs like a gazelle; a cheetah is agile and capable and moves quickly, gracefully, forcefully.

So why did Aristotle and others suggest that the unmoved mover is the highest state? It's because movement was tied to a dualism, a concept now mostly outdated thanks to our understanding of evolution and the normal functions of the nervous system. Movement was thus seen as a kind of privation: less mind and more body. Nevertheless, Aristotle thought like a biologist, capturing and cataloguing living things. Indeed, as human beings, we do a lot of cataloguing, and we do it quite easily because from a young age we are generally taught to navigate the world through categories. Thus we are quick to classify objects in terms of the kinds of things we believe they are.

Much of our categorization is tied to action, with sport often considered the quintessential moment of action. But sport is not just action; there is no dualism of mind and body. Whatever one means by "mind" (of which there is no univocal definition), sport is rich in it. And running through sport is thought—embodied thought (Varela, Thompson, and Rosch 1991), structured and practiced through well-worked habits (Peirce 1878).

Movement is replete with thought. While movement is not the essence of sport, it is a fundamental feature of it: think of the swing of a baseball bat, the bunching and unbunching muscles of the sprinter's calves, the tension in the still moment before a golfer sinks the ball in the hole, the alignment of the body in a single plane as the archer pulls back on the bow. All this is movement, but it is movement organized and driven by thought, purpose, and intent. And while sport always involves movement, not all movement is sport.

In this chapter, I begin with a context for understanding sport, given our capabilities and our cultural evolution: sport lies in the continuous fluidity between biology and culture.

WHAT IS SPORT?

When is movement considered a sport? People tend generally to agree on what is sport and what is not and even to share the same sense of ambiguity about the classification of certain activities. Rather as Supreme Court Justice Potter Stewart said of obscenity in 1964, we know it when we see it. But there is no one feature that defines sport, except perhaps that it is competitive and that there are always winners and losers. Even in elementary-school athletics where everyone goes home with a trophy, the children are always—and sometimes cruelly— aware of who won and who lost. But sport is also a type of game, and Wittgenstein (1953) and later game theorists have identified a family of properties that loosely defines a game. One feature games have in common is that there are rules to be followed and a language or logic to understanding the activities and participating in them. There may be little in common between skiing and soccer, between snowboarding and hockey, but they all are rule based and have winners and losers based on those rules. Indeed, a signal feature of those activities that some people would prefer not to regard as true sports (although no one doubts that their practitioners are true athletes), activities such as figure skating, is that they contain subjective elements such as "artistry" that can be decided only arbitrarily by judges and not by a strict, clear application of the rules.

While all sports are a type of game, not all games are sports. Chess and Scrabble are conventionally considered to be games but not sports, even though they can be played competitively and there are clear winners and losers.

One reasonable view about how to distinguish play from sport (Guttmann 1978) takes the list of all things we consider to be play and pares it down into games, contests, and then sports. Play includes spontaneous play and organized play (games). Games then can be divided into noncompetitive games and competitive games (contests). Finally, contests can then be categorized by intellectual or physical contests, and the latter type of contest is what we consider to be sport.

To be truly considered a sport, an activity's practitioners must exhibit some sort of physical prowess, some feat of practiced, intentional, applied movement that reaches beyond the ordinary. They must be athletes.

THE HISTORY OF SPORT IN A NUTSHELL

We know games are very old. What appear to be gaming boards first surface in the Neolithic period. It's hard to believe that sport does not have an even older history. Archaeological evidence suggests that wrestling is the first publicly expressed, original sport; this is not surprising, since rough-and-tumble play is a core feature in the lives of many mammals and certainly of primates.

The first truly organized sporting events seem to have been competitions associated with Greek religious festivals honoring gods, commemorating mythical events, and marking seasonal changes, starting about 2,500 years ago (Guttmann 1978, 1991, 1992, 2004). The origin of the ancient Olympic Games is speculative, but they may have begun as a commemoration of Zeus's defeat of Kronos in a wrestling match.

The original Olympic Games are merely the best known of such festivals (Guttmann 1992), which were held throughout ancient Greece. Other notable examples of athletic festivals include the Pythian (582 BC) and the Nemean (579 BC) Games that were held to celebrate Apollo as well as the Isthmian Games to celebrate Poseidon (582 BC) (Guttmann 1978, 2004). As well as seasonal festivals, sporting events were associated with life-cycle markers, especially funerals. The *Iliad*, for instance, describes the funeral games that Achilles held in honor of his friend Patroclus, who was killed in the battle for Troy. The actual sports played at such festivals were many, including footraces of different lengths, boxing and wrestling events, horse races, and the pentathlon (Guttman 1978, 2004).

The ritual aspects of such games—the competition and the spectacle—must have fueled excitement. Ancient Greek audiences were interested in form and beauty and excellence. Though statistics emerged only in the seventeenth century within the context of state records (births, marriages, deaths, etc.,), keeping records of sporting events, perhaps noting times and comparing trends, and certainly following the successes and failures of specific athletes, goes back to the ancient Olympics. The emphasis was on form, not experiment, and certainly not probability. The organization of numbers is infused in all thinking, and it is one of the primary ways we keep track of events.

BALL GAMES

Balls, made of rubber in Mesoamerica and of leather and textiles in other parts of the world, feature in ancient games. Pre-Columbian ball courts in the Americas survive at a number of sites, and the balls and games are depicted in detail in Mesoamerican art (Stone 2002). Indeed, many pre-Columbian ball games in Latin America are believed to have involved the human sacrifice of the losers or possibly of the winners (as the most fitting gift to the gods). Balls are also central to many loosely organized community-based games, for example, Scotland's ba' game (see below), in which local friendly and not-so-friendly rivalries are hashed out.

As a New York City child I played a lot of such casual ball games with the other neighborhood kids: punch ball, stick ball, throwing the ball off the stoop. Little did I know in what a long tradition I was participating. Stick ball, which involved hitting a rubber ball in the street, with marked sewers as goals, was an after-school game played between cars. But beyond the fun, kinship relationships, neighborhood rivalries, and

FIGURE 1.1

North American Indians played a diverse set of games and sports. Lacrosse, a Native American invention originating in upstate New York, is still played today.

Source: Culin (1907).

social hierarchies were also being negotiated. Formal team sports are an outgrowth of this sort of casual play, and the larger the group, the larger the terrain.

And sometimes the terrain in sports gets very big indeed. Sometimes it is a proxy for war. Many anthropologists believe that sports are a kind of ritualized war, which is in itself a form of competition with winners and losers. I mentioned above the high stakes involved in pre-Columbian ball games. Aristotle referred to sports as an expression of aggression without the blood—or at least the amount of blood shed in warfare. Diverse contact sports involve aggression, and blood is shed by both the fans and the players (think of Manchester United's matches). That is the nature of some sports: brute aggression and group solidarity.

Early on, some sense of what we would now call nationalism, with devotion to a particular team or competitor based on their geographic or ethnic origins, was introduced into sport. It is that nationalism or regionalism (Argentina versus Brazil, Oakland versus San Francisco) that leads to the roar of identification and glorification that goes up in any modern sporting arena. I marveled at the Cuban boxers and the Eastern European gymnasts—their dedication and the fanatical followings they acquired—during the heyday of the communist countries. Again at the 2014 Olympics in Sochi, we saw sport stand in for deadlier rivalries on the world stage. The ritual significance of the earliest games is surely alive in modern sport.

Such identity-based purposes for games have an old and rich history. In the Orkney Islands, off the northern tip of Scotland, the Kirkwall ba' game is played in several iterations over the Christmas and New Year holiday, traditionally between the Uppies (the inhabitants of the upper part of the town) and the Doonies (those who live around the harbor). This is an exuberant and sometimes vicious free-for-all with few rules apart from getting the ball across the dividing line between the two parts of town, and the scrums can number up to 350 men (women and boys do not play in the main game, but they may have their own set-to). While Kirkwall ba' goes back at least to the seventeenth century, it may be far older. The presence of similar games is attested to in many parts of the British Isles and America. If sport began as ritualized war, designed to bleed off aggression between natural rivals, games such as the Kirkwall ba' game are the closest modern relative of that ancient sport.

The cultural evolution of association eventually led to small groups coming together to form organizations (de Tocqueville 1848). This was most apparent starting in the 1850s in the United Kingdom. Clubs, an important feature of life in the Victorian era, were organizations for social interaction (Szymanski 2006). Modern sport is essentially secular and tied to records (Guttmann 1978). This timeline of sport reflects both aristocrats and commoners.

Associative sports in Britain and the United States paralleled the development of public societies, government bodies, and other social organizations. The formation of cricket, golf, and horse-racing clubs marked the beginnings of early sporting organizations that would one day become the modern franchises we now associate with games such as football, basketball, and baseball (Syzmanski 2006).

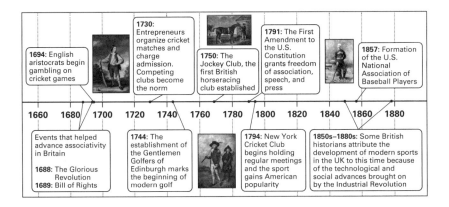

FIGURE 1.2

A timeline depicting how association—in this sense, the formation of sport specific clubs—led to the development of modern sports.

Source: Adapted from Szymanski (2006).

Cricket, a precursor of baseball that still has a fanatical following in England, Australia, India, and Pakistan, is a prime example of associative sport. Cricket clubs were common from the mid-nineteenth century through the mid-twentieth; every English village once had a cricket club and could field a team. Horse-racing and jockey clubs, more popular with the upper classes, were also common during that period.

As Syzmanski (2006) points out, associations also developed elsewhere in Europe, such as France and Germany, and some were tied not only to games and sports but also to education and to the gymnasium as a place to express and develop athletic excellence.

THE ORIGINS OF SOME SPORTS

It is more difficult to find the exact origins of many sports played by individuals because leagues and chapters were not established for them in the same way they were for team sports. Boxing may have emerged as a sport

in the Greek Olympics of 688 BC. It evolved into prizefights in sixteenth-
to eighteenth-century England, eventually becoming the sport we know
today. Cycling as a sport began in France shortly after the invention of
the bicycle.

Golf appeared in many varieties throughout Europe in the Middle
Ages, but it developed most thoroughly in Scotland. Gymnastics and
acrobatics have evolved over seven thousand years in many places, both
as forms of entertainment and for military training. Training centers for
civilian gymnastics opened broadly in many places in Germany in the
nineteenth century (Mandell 1999).

The origins of these sports are diverse. Baseball developed out of eighteenth-century folk games in England, and lacrosse is a modification of
a game played by Eastern Woodland Native Americans and some Plains
Indians tribes in Canada. Basketball did not evolve on its own, as many
games probably did, but was invented in Massachusetts in 1891 by the
physical-education teacher Dr. James Naismith, the son of two Scottish
immigrants to Canada. He invented this indoor game to provide exercise
for a YMCA class of nonathletes (Naismith 1941). Similarly, volleyball
was invented only four years later (and just up the road from basketball's
birthplace) by William Morgan (Dearing 2007).

The origins of soccer, on the other hand, are unclear. While in the
second and third centuries BC a similar game was played in China, the
origin of the modern game is thought to have emerged in fourth-century
England (Goldblatt 2008). Rugby and football descend from similar versions of those games played in nineteenth-century England (Goldblatt
2008). Hockey has been developing for close to four thousand years:
Egyptians, Ethiopians, Romans, Greeks, Aztecs, English, Irish, and Scots
all played some version of it (Reddy 2011).

One sport that may date back *thirty* thousand years is wrestling. Such
events are tied to diverse forms of adaptation. Violence was common, and,
as Hobbes (1651) says, life was "nasty, brutish and short." Humans have
probably always found violence entertaining. We have only to think of the
excesses of gladiatorial combat or modern racing enthusiasts' enjoyment
of Formula 1 and NASCAR crashes to realize how fundamental violence
is to our fondness for sport. Hockey, football, lacrosse, and boxing also
exhibit a great deal of violence.

FIGURE 1.3

"The Boxing Boys." Frescos from Thera (modern-day Santorini), ca. 1600 BC.

Source: The Thera Foundation (Murray 2010).

THE MODERN OLYMPICS

It was at the end of the nineteenth century that the modern Olympics was resurrected and placed in the context of the Hellenic spirit of excellence and education (Coubertin 1896). The five linked rings in the logo of the Olympics represent the five continents united by sport (Guttmann 1978, 1992, 2004). At a time of revolution and change, an ancient symbol of the athlete was resurrected in the form of the classical age.

But even as we acknowledge the higher aims of sport—health, character, ambition, self-control, international unity—the ancient ritual still lurks beneath it. One has only to read the news coverage of the 2014 Olympics in Sochi, for example, vaunting the successes of American national sports figures and quietly (or not so quietly) enjoying the minor disasters of Russia's handling of the games, to see that in some ways sport is still sublimated war.

We bond together in these displays of "us versus them" in our winning and their losing, whether it is cockfighting or human violence: our team, our group. A common theme in the understanding of vicarious participation in violence (Aristotle 1968, James 1887, Lorenz 1981) is some sort of release factor for the one doing the watching: the fan (Goldstein 1998).

But besides violence, aesthetic appreciation and even the erotic are essential to sport (Guttmann 1996). The beauty of the body—its form and muscle and excellence of movement—is as important to spectators as anything in sport. Force and strength are elements of both beauty and violence in many activities: control, balance, strength, and lightness, a balance of contrast: a glorified mean.

This movement is on land and sea; the boats we build are for exploration as well as sport. The utter continuity in sport of cephalic capability and demonstrable utilization and expansion is transparent. Sport is a spectacular window into evolution. Judo and other martial arts are tied to disciplined physicality; spiritual expression and physical prowess are expressed in movement, control, and action. Bursts of speed and power balance against lightness and elegance: butterflies and bees.

BEAUTY AND SPORT

Athleticism has the same uncanny ability to awaken our aesthetic impulses and make us ignore the sweat and carnage involved (Curley and Keverne 2005). Imagine the sensorimotor capabilities of Michael Jordan, predicting where his teammates are going to be, rich in predictive and regulatory appraisal systems. The extraordinary balletic quality he brought to the game at his best is on par with the elegance and sophistication of the dancing of Dame Margot Fontaine.

Indeed, athleticism is rich in beauty (Gumbrecht 2006). The aesthetics of sport goes back to the classical world; there is a reason athletes feature so prominently in its frescoes and on its vases and survive as statues. The sense of beauty in the expression of the body in sport homes in on and elevates it to its highest levels. Many of these athletes and superstars function within contexts of adoring publics, across towns, cities, states, countries. And the burdens on the athletes are high.

We respond, as we do to shining ranks of soldiers marching elegantly to their deaths, to the sheer perfection of physicality. But that physical perfection begins not in a shapely set of thighs or powerful shoulders or a muscular core. It starts, really, in the brain. Thus, while competition, play, social contact, training, and winning are key elements of sport, mental excellence is the feature that decides whether one gets there.

2

SPORTS, BRAIN, BODY, AND THE WORLD

Sport as a practice and cognitive event is largely action oriented: there is thought in action, even if it is not particularly conscious thought. Habits codified in the motor regions of the brain underlie action (Jeannerod 1997). Muscle memory is fundamental; it is a metaphor for the organization of action.

Throughout sport runs the experience of knowing and doing at the same time, but as any athlete will tell you, this works only when not done to excess. Becoming frozen in thought or immobilized by overthinking is ineffective in terms of action response. Sport reveals the endless continuity of action, perception, and predictive expectations, three things essential to the organization of action (Clark 1998).

Think, for instance, of how much easier it is to drive a car or ride a bicycle once you have mastered the art of it; instead of having to think consciously about several skills at once, you are able to let your body do a certain amount of the work on what seems to be autopilot. In the same way, many of the cognitive and adaptive achievements in sports are unconscious. So when your conscious brain or your coach yells out, "Don't think so much," you know what that means.

We are anchored to others in both a social and ecological context, and this context anchors action and harbors memory (Gibson 1979, Clark 1998). Memory concerning how to perform when put in a familiar terrain associated with an action is not uncommon in the athletic world. The act of performance, or what Andy Clark calls "being there," brings the mind, brain, and world together as one. Sport is a clear reminder that "Descartes' error" (Damasio 1996, Lakoff and Johnson 1999) was not only the mistake of the separation of a mind from a body but the larger (and common) mistake of the separation of the social and ecological worlds we live in and are constantly adapting to through cephalic function.

The word *cephalic*, for me, is a way of referring to mind/brain together, the two embodied in action. Moreover, the continuous function of anchoring memory into social context is a way of understanding adaptation; there is so much to remember, so to reduce the burden on neural function, we ritualize (Donald 2004). Social context exists in most of what we do, and it exists especially in sport.

Sport reminds us why it is as important to not exaggerate the mental as it is to remember the mental's role. Practice, practice, and more practice within learning the form is essential to the development of excellence in sport; shooting for the hoop over and over again, or launching into a triple salchow jump fifteen times a day, so much so that it becomes embedded in one's head. Stephen Curry of the National Basketball Association's (NBA) Golden State Warriors is a prime example of how practice is essential to the development of excellence in basketball. Curry is renowned for his practice, practice, and more practice. And, he expresses genius (Davis 2015). The continuous detection and computation of angle and terrain became one in a moment of performance.

In this chapter I will provide a sense of the brain, in particular the motor regions that underlie diverse sport, and will discuss the diverse information molecules across the brain that underlie our capabilities. We came into the world prepared for sports, and sports continues and accelerates our sense of physical excellence and beauty.

NEURAL DESIGN AND MOTOR CORTEX

Design principles are efficiency oriented to reflect our species and its capabilities for survival in local niches. The neural systems reflect specificity, separation, minimization, energy costs, size convergence and divergence, redundancy (80 percent of neocortical neurons are pyramidal cells), speed, and accuracy in neural function (Sterling and Laughlin 2015, Laughlin 2001).

The neural speed across different neuronal systems depends on physical size and shape (Sterling and Laughlin 2015). Of course, theory and observation infuse the seeing of pyramidal cells (Hanson 1971, Swanson 2003);

seeing is in context, in the lens of a microscope, or through neuroscientific techniques. But no matter how small or abstract the concept of pyramidal cells seem, their existence and importance is real enough in their application to one's ability to participate in play, sports, and war.

The motor system is a large part of the brain. And within the motor system there are appraisal systems that are both affectively laden and computational; movement is couched within the appraisal sensibility vital for survival. Diverse neural/behavioral systems reflect the niche that one is coping with and surviving in, including the practice of sport.

Mapping of the motor cortex has a long history (Gross 1998), both in medicine and in art: medical illustrators have been depicting the motor

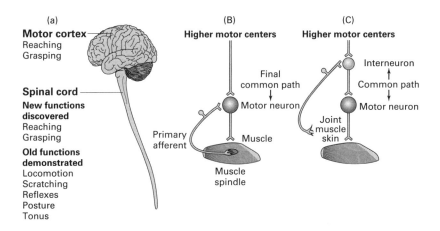

FIGURE 2.1

Motor principles.

(a) Skilled voluntary movements such as reaching and grasping are commonly considered to be dependent on neural circuits mainly located in the motor cortex, whereas stereotyped motor behavior such as locomotion, scratching, reflexes, posture, and tonus are controlled by spinal circuits. However, as described in this review, the descending command for skilled reaching and grasping can also be controlled by spinal circuits. (b) Schematic neural diagram showing monosynaptic input from muscle spindle Ia afferents and descending pathways from higher motor centers in the cortex and brain stem. Sherrington (1906) considered motoneurons to be the "final common path." (c) Schematic neural diagram showing disynaptic input, via an interneuron, from afferents and higher motor centers to motoneurons.

Source: Alstermark and Isa (2012).

cortex over many centuries (Finger 1994). We now know that diverse neuronal projections run from the primary motor cortex to the spinal cord,
shoulder, finger, and elbow muscles (Rathelot and Strick 2009). Their
specific capabilities include the expression, related directly to muscle, of
highly evolved and increased flexibility for skill development. These events
are expressed in the brain, the final common factor in motor expression.
The development and expression of motor skills is thus controlled by the
primary motor cortex (Matsuzaka, Picard, and Strick 2007).

Amazingly, the primary motor cortex projects not only to muscle
(Kakei, Hoffman, and Strick 1999) but also to diverse peripheral organs,
including the kidney. Over the last thirty years we have discovered that
the central nervous system and cortical sites have direct projections to
peripheral organs themselves (Swanson 2000). Muscle memory in sport
stretches across a distributed neural system.

Once we understand the importance of the motor cortex to physical
capacity, we can understand that physical capabilities are tied to cortical expansion, a topic we'll discuss in more detail later in the context of
evolution. For now, just consider the expanded range of primate primary
cortices (Kaas 2013).

As neuroscience has developed, the very idea of the motor cortex has
changed, and the once sharp separation between cognitive and motor
expression in the cortex has been diluted. In addition, the sharp separation between strictly motor and sensory systems from cognitive and
computational systems now appears to be far more porous. Of course,
that does not mean that one can detect specific regions of the brain tied
to different regions of motor expression in the organization of action and
different regions of both primary motor and premotor regions (Dum and
Strick 2013, Rizzolatti and Luppino 2001), but it does make it easier to
understand how athletes can do what they do. These are motors that go
literally from cortex to peripheral muscle; practice and underlying cognitive and physical capability exist thanks to rapid neural connectivity.

The evolved motor cortex is involved in the adjudication of our sports
experiences; after all, the very possibility and existence of sports and
games reflect the cortex that we indeed have. For instance, it is not clear
what the region known as the premotor cortex does other than taking
care of the planning side of motor expression. The premotor expectations

are continuous with the motor-cortical control and the speed that makes LeBron James' performance possible. Perhaps this is a degree phenomenon. Some parts of the motor cortex are directly tied to motor expression as we know it, but premotor regions are less tied to planned action. This distinction, though, is on a continuum and not absolute.

But in regard to how motor systems were understood in the nineteenth and most of the twentieth centuries, the dubious and pernicious distinction between thought and action has been undercut. The boundaries between the premotor and the motor regions are porous, as are the functional relationships in the organization of action. And if it is about anything, sport is about action, at least in the common sense of the term.

Moreover, motor areas of the brain (for example, Broca's) are always tied to cognitive systems. After all, what is more central in the pantheon of cognitive systems than language or syntax (Pinker 1994)? Computational systems that parse syntactical linguistic expression are fundamental to our brain for learning to become anchored with others in our social world. And diverse kinds of neural generators orchestrate behavioral patterns; two regions of the brain that facilitate this are Broca's area and the basal ganglia (Berridge 2004).

Regions of the brain such as Broca's are tied to other forms of syntactical relationships besides language (for example, musicality), and the organization of syntactical expressions is revealed in what Lashley (1951) called the serial order of behavior. Neurotransmitters such as dopamine are essential in the serial order of basic motor movement as well as in the syntactical relationships in language (Berridge 2004, Ullman 2004), and they are essential in learning and in embodying the rules of sport.

CEREBELLUM

Another region of the brain significant to our understanding of motor control is the cerebellum, which sits on the brainstem and is separate from the rest of the brain. It has long been thought to be linked to the expression of fine motor movement. Human cortical expansion is particularly marked in the cerebellum. After controlling for brain volume, expansion has been estimated to be 40 to 50 percent larger in apes than in monkeys

(Rilling and Insel 1998). Cerebellum volume, which is tied to motor control, increases dramatically in humans compared to nonhuman primates and monkeys (Rilling and Insel 1998).

The cerebellum seems to underlie motor control. Regions of the cerebellum receive projections from the motor cortex (Coffman, Dum, and Strick 2011), and the cerebellum is knotted to diverse cognitive functions. Indeed, cerebellar connectivity to the basal ganglia further implicates a third anatomical structure (Bostan, Dum, and Strick 2013). The basal ganglia, long known to be tied to motor control, differs in appearance between species; some species (for example, birds) have an expanded basal ganglia (Reiner, Medina, and Veenman 1998).

NEUROTRANSMITTERS AND THE ORGANIZATION OF ACTION AND THOUGHT

Neurotransmitters tend to be broad in terms of regulation. We've talked a bit about dopamine and its role; another neurotransmitter in the brain is gamma-aminobutyric acid (GABA), an important inhibitory neuron colocalized with a number of other transmitters (Swanson 2000, 2003).

TABLE 2.1 Common neurotransmitters in the CNS

Catecholamines
Dopamine
Norepinephrine
Epinephrine
Indoles
Serotonin
Melatonin
Cholinergic
Acetylcholine
Amino acids
γ-aminobutyric acid (GABA)
Glutamate
Aspartate

Information molecules are diverse and overlapping, and they display divergent evolutionary trends. Dopamine, for instance, stretches back half a billion years; the monoamines are also no newcomers to the planet. Dopamine is part of the larger class of monoamines and indolamines (a class that also includes serotonin).

The basic structure of catecholamines, like many of the information molecules, is found in both vertebrates and invertebrates (Yamamoto and Vernier 2011). Information molecules in peripheral systems are also produced in the brain. In the periphery they play diverse regulatory physiological functions; in the brain they play more integrative functions. In the periphery they are regulative (dopamine, corticotropin-releasing hormone), and in the brain they are physiological.

For instance, in boxing we think the neurotransmitter norepinephrine would be operative for attention, dopamine in stamina and maintenance, and serotonin in pace, tone, and timing. Of course, some of this is speculation. But a number of studies show attention is tied to norepinephrine in the brain, dopamine is vital in sustaining action, and serotonin affects temperament and attitude.

Hormonal messengers are fundamental information molecules, and there are many kinds. Dopamine and norepinephrine or serotonin, when expressed in peripheral tissues such as the adrenal gland, are thought of as hormones; in the brain they are neurotransmitters. They are the same molecules: depending on where they are expressed, they can be involved in the organization of behavior or be physiological signaling systems (Herbert and Schulkin 2002).

Basic neurotransmitters such as dopamine play diverse roles in the regulation of brain and body. Dopamine is known to be linked to Parkinson's disease, a central nervous disorder of progressive devolution of motor control, resulting in the inability to regulate movement and eventually progressing from a state of constant shaking to being unable to move, frozen in place. Thus dopamine is usually thought of in the context of movement.

Muhammad Ali, a boxer of fantastic skill and real beauty, is a well-known example of a person with Parkinson's disease. I was lucky enough to see him on Fifty-Seventh Street in New York City many years ago, trying with shaking hands to pick up a child. That simple task was hard

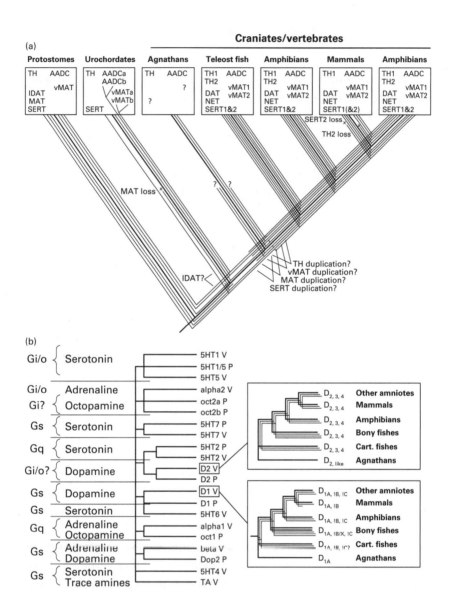

FIGURE 2.2

Evolution of the molecular components of the monoaminergic systems in chordates. (a) Protochordates have all the basic molecular components of the monoamine pathways found in vertebrates. In urochordates, both MAT and iDAT do not exist, but since MAT is present in amphioxus, it may have been lost specifically in urochordates. The loss of iDAT may have occurred earlier since it is not present in amphioxus. In vertebrates, MAT has been duplicated to provide DAT and NET, which are not thus orthologous to invertebrate iDAT but to invertebrate MAT. Similarly, SERT has been duplicated in jawed vertebrates. vMAT and AADC have been specifically duplicated for him to complete, indicating the beginning of a disease that would progressively result in the devolution of function.

But movement and cognitive systems are endlessly overlapping, so it makes no sense to refer to motor systems as noncognitive. Indeed, the prediction of reward is tied to the activation of "the basal ganglia, a repository of dopamine innervation and activity" (O'Doherty et al. 2004). Some of these regions (Broca's and regions of the basal ganglia) are linked to the syntactical function of human language (P. Lieberman 2002; Ullman 2001, 2004). Basal ganglia function is linked to statistical computational competence (Knowlton, Mangels, and Squire 1996), including events that are affect laden (Saper 1995; Berridge 1996, 2004) and that underlie the diverse forms of sports.

---

in the urochordate lineage. Close to the emergence of vertebrates, TH, AADC, and vMAT have been duplicated. But since no genomic data are available yet in agnathans, it is currently impossible to know if this duplication took place before or after the emergence of jawed vertebrates. TH2 and SERT2 have been lost in placental mammals. Abbreviations: AADC, aromatic amino acid decarboxylase; iDAT and DAT, invertebrate form (i), and vertebrate form of dopamine transporter; MAT, monoamine transporter; NET, noradrenaline transporter; SERT, serotonine transporter; TH, tyrosine hydroxylase; vMAT, vesicular monoamine transporter. (b) The molecular phylogeny of monoamine receptors in bilaterian animals reveals that most classes of monoamine receptors predated the origin of chordates and vertebrates. Classes of orthologous receptors in vertebrates and protostomes (most sequences come from ecdysozoan insects and nematodes) are transducing signals in cells via the same G protein (question marks correspond to the cases when the nature of G protein is not known), highlighting one of the major constraints on the conservation of the receptor sequences throughout bilaterian evolution. For example, $\alpha 1$ adrenergic receptors are orthologous to octopamine 1 receptors (Oct1) and $\alpha 2$ adrenergic receptors are orthologous to octopamine 2 receptors (Oct2), but both D1-like and D2-like receptors are also dopaminergic in protostomes. The topology of the tree also shows that receptor classes that bind the same natural ligand (e.g., DA) are not grouped together, suggesting that each class of receptor acquired independently and convergently the ability to bind a given neurotransmitter. Inside the rectangle, a simplified version of the phylogenetical relationships of the D1 and D2 receptor is presented. Three subtypes of receptor exist in each class, with the notable exception of mammals.

Source: Adapted from Yamamoto and Vernier (2011).

For example, one fMRI study linked the valuation of others to social circumstance and comfort of living, and it found greater activation of the ventral striatum (Fliessbach et al. 2007). In diverse contexts of learning, tasks tied to reward have been connected to the human basal ganglia (O'Doherty et al. 2004; Squire, Knowlton, and Musen 1993). In one such context, using fMRI, several regions of the striatum can be shown to be tied to the learning of a task linked to reward, even in those who are deficient in the learning of simple tasks (Schonberg et al. 2012).

From an adaptive social point of view, the expansion of basal-ganglia function in primates is linked to learning systems and prediction of events (skills essential in many sports), with one set of dopamine-related neurons tied to uncertainty and the other tied to expectation of outcomes. In both cases, the anticipation of events is part of allostatic regulation, less on the reflexive response to change and more on the anticipatory systems predicting change to come; this predictive capability is essential in the formation of social groups, social cohesion, and competition for resources.

Dopamine is the one neurotransmitter linked to reward (S. Wise 1985; R. Wise 2005; Wise and Rompre 1989; Koob et al. 1994; Schultz 2002, 2007). Dopamine is also fundamental to the processes that underlie both

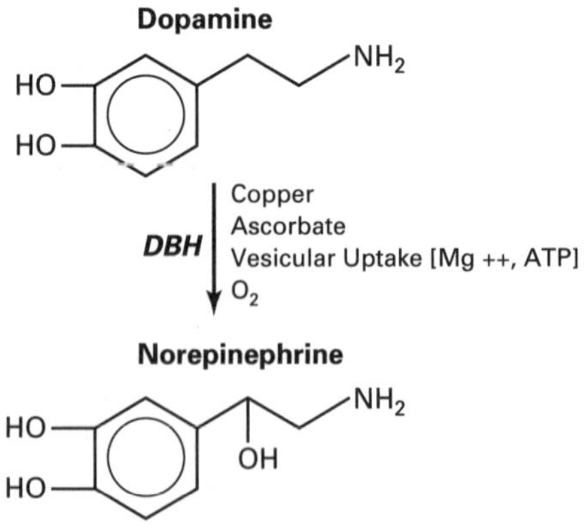

FIGURE 2.3

Dopamine and norepinephrine.

the organization of action and the organization of thought. And reward is tied to motivation; motivation requires attention to the external world for needed resources and the organization of action to acquire the desired objects. But the concept of reward is complicated; it is not simply one act, and an attempt to define the term becomes circular (S. Wise 1985; Gallistel 1980, 1990). Nevertheless, to understand sport, you must appreciate the role of dopamine.

Dopamine is produced in the brainstem and dopaminergic neurons project to various brain regions, including the basal ganglia and motor regions of the cortex. This neurotransmitter is also produced in peripheral systems such as the adrenal gland, the same gland that produces the adrenal steroid cortisol. Dopamine and cortisol both play a part in the mobilization of action.

Again, dopamine is tied to basic motor control and also to statistical inference (Schultz 2002). This is essential in sport, particularly social sports in which predicting the behavior of others is key (for example, football). One set of dopamine systems, for instance, is activated when expectations are disrupted and new focus is required to facilitate expectations. Dopamine is also knotted to incentive salience and objects of relevance. Incentives are the things that matter, or are thought to matter, in the organization of action and the mobilization of behavior (Berridge 2004). Sport is filled with such events in the context of competition.

Dopamine is a primary neurotransmitter in the brain, as is norepinephrine. Epinephrine is found solely in adrenal modular cells (Goldstein 2000). A conversion process from tyrosine is fundamental in the production of dopamine. Dopamine is also a precursor in the metabolic conversion to norepinephrine; one is essential for the organization of action and cognitive expectation, the other for attention (Aston-Jones and Cohen 2005).

Dopamine is involved in both action and thought; as such, it is a necessary chemical for us in maintaining a coherent world in which to function. The dopamine pathways in the brain underlie a number of behavioral functions that range from syntax production and probability reasoning to the activation and learning of specific motor programs. For example, Parkinsonian patients (who have depleted dopamine levels in the brain) have trouble with ordinary syntax as well as with motor control. The mechanisms for generating action are not separate from the mechanisms for thought; the serial order of behavior is organized by a number of regions of the brain, including the basal ganglia, in which dopamine is a primary neurotransmitter in the organization of behavior.

Dopamine neurons are linked to the learning phase in a number of paradigms and to the response to novel events (Schultz 2002). Dopamine neurons in the frontal cortex and striatum are active in anticipation of these events (for example, gustatory, auditory, and visual rewards).

Dopamine is not simply a neurotransmitter underlying the brain mechanisms linked to reward. It is much more complex—even when dopamine is blocked, animals can still "like" things (for example, sucrose). Indeed, dopamine is more tightly linked to the motivational component of pleasure-related events. It can be separated from the predictive reward components, and some endorphins are linked to the ingestion of a reward (Berridge 2004). Thus, importantly, dopamine is not only essential for the organization of drive, of wanting something, but also for the salience of events.

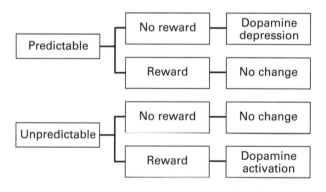

FIGURE 2.4

Predictable and unpredictable events and dopamine.

Source: Adapted from Schultz (2002).

The dopamine system does not function in isolation. Dopamine is colocalized with a number of other neurotransmitters and neuropeptides at a variety of sites in the brain. Neurons in the nucleus accumbens receive input from the hippocampus as well as from the brainstem (serotonin neurons and glutamate neurons), the cortex, and the thalamus. Neurons within the nucleus accumbens contain receptors for many other neurotransmitters.

In the context of the organization of action, dopamine is just one neurotransmitter among several. What is interesting is that this one neurotransmitter, which is essential in the organization of movement, is also essential in the organization of thought—namely the prediction of events—which is crucial in sports (and a wide variety of our other activities).

COGNITION/MOVEMENT AND SPORT

Crawling, walking, and running, intermixed with speed and the ability to control those movements to facilitate exploration, are important milestones in our development. We must continually make sense of the world around us, navigating a space full of physical objects to be mapped and understood (Kagan 1984, Carey 1985).

We track events and fill in details. We start out with a rudimentary sense of mechanical relationships in infancy, which we refine as we grow and develop (Keil 1989). The brain's maturation process is particularly dramatic in utero, but it continues at least until the end of the teens and, as new research shows, in some respects through our mid-twenties. One issue is that the formation of neural connections coexists within the milieu of epigenetic experiences throughout life. This matters for all human activities, including sport.

Indeed, we come ready and prepared with orientations for exploration, for play and sport. Moreover, organ-related functions appear quite early in the young's lexicon of understanding. While they may not yet be scientists, three- and four-year-old children catalogue events into categories relevant to them: animate and inanimate objects, familiar objects, reliable events, faces that have meaning, food, etc. (Carey 1985).

We come prepared to recognize animate objects, causal relations, and self-propelled objects (Premack 1990). These are all vital skills for survival, and they also underlie the action in sport. Along with our developing cognition of our physical surroundings, we very early on learn to detect the intentions of others by bodily and—particularly—facial cues. A "theory of mind" (Premack 1990) inherent in human and perhaps some other big-brained animals underlies a vast array of rapidly produced inferences about others.

This capability, which appears early in development, is tied to the prediction of others and is knotted to pretend play and imaginary objects (Peterson and Wellman 2009). Fooling others, play-pretending, and misleading others' expectations are well within the capabilities of the very young (Gopnik and Meltzoff 1993). All these cognitive skills have a role in sport.

IMAGINING ACTION AND ACTING

Both sensory and motor systems contain mechanisms to appraise events, and both are embedded in the organization of action and in the anticipation of events. In sports, the brain is rarely if ever a passive organ; predictive expectations reflect neural design and, perhaps, particularly cortical design.

The motor system in learning and playing is dispersed across a wide array of neural sites and systems in the brain, and these sites and systems are responsible for diverse functions, such as the learning of the different elements of a new motor task and then the consolidation of that task (Passingham 2008). Regions of the neocortex, motor cortex, premotor cortex, basal ganglia, and cerebellum underlie these capabilities.

But imagining and observing action, as well as practicing that action, enhance the skills of athletes and dancers—and the rest of us too, within the bounds of reason (Beilock and Gonso 2008). The capacity for imagination enhances action. As it turns out, motor regions are active in both imagining an action and in the performing that same motor movement (Decety 1996). Sport is rich in imagining action.

MOTOR REGIONS OF THE BRAIN AND TOOL USE

Many species use tools—defined as an object modified for use in achieving a specific goal—in adapting to their environment. Tools and tool use reflect the evolution of cortical and subcortical systems in the brain that participate importantly in tool use, tool making, and tool recognition (Gibson and Ingold 1993, Johnson-Frey 2003, Martin 2007, Martin 2015). Tool use is an expression of an expanding cortical motor system in which cognitive systems are endemic to motor systems (Martin 2007, Ullman 2004, Lieberman 2002). The tools we use are extensions of our bodily sensibilities: consider the baseball bat or the use of the stick in hockey and lacrosse. As when we play a musical instrument, these tools become part of us as we use them in sport.

The use of tools in sport emerges from the same basic capabilities as does other tool use. Moreover, regions of the brain are prepared to recognize differences between different kinds of objects, one of which is mechanical tools (Martin 2007). Importantly, frontal motor regions have been linked to the motor features of tool use (Johnson-Frey 2003, Johnson-Frey et al. 2003). Like acting and the imagination of action, the regions of the brain activated simply by the *sight* of tools and by the *use* of tools are the same (Jeannerod 1997, 1999).

The expansion of cephalic function that underlies tool use serves both physiological and behavioral regulation. An expanded motor system with diverse cognitive capacities was no doubt pivotal in our evolutionary ascent. It is not just the evolution of the cortex or brain that is knotted to social function; all tool use and other diverse abilities are tied to the fact that we are social animals.

The nineteenth-century American philosopher and psychologist William James understood in particular the inherent biological embodiment of cognition in motor control and expression. Cognitive systems are pervasive in action, and they are not detached systems but embodied systems in habits and action.

Basic motor reflexes at the level of the brainstem are orchestrated by higher cortical functions (Sherrington 1906). But the notion of a reflex arc

(Dewey 1896) is too narrow for understanding human and, more generally, mammalian behaviors, let alone sport; reflexes bend into choice in the expression of behavior in general and sport in particular. Regions of the brain such as the motor cortex and basal ganglia are tied to forms of behavioral expression. Think, for instance, about the diverse forms of creative movement in the basketball playing of Michael Jordan, blending form and function and creative diversity of expression.

Indeed, syntactical conditions are not just a feature of language but also of movement (Berridge 2004, Lashley 1951). The organization of action is a mixture of reflexes and choice, creative expression and rock-solid habit, and reflexive expression (Gallistel 1980). The important insight is that syntax is embedded in sport movement and expression. Broca's region of the frontal cortex and regions of the basal ganglia, for instance, are tied to syntax and underlie language (Ullman 2001). The organization of thought and movement is further revealed by statistical organization and in the prediction of events; regions of the basal ganglia underlie statistical inferences, which are so essential both in the organization of action and in sport. The basal ganglia is fairly constant across mammals, birds, reptiles, and amphibians (Reiner, Medina, and Veenman 1998; Smeets et al. 2006). And the basal ganglia, which is tied to the organization of movement, is rich in neurotransmitters and neuropeptide expression and receptor sites.

CONCLUSION

The motor systems in primate brains tell the story of the evolutionary changes that underlie our capability. Moreover, while we do not have to move to think, appraisal systems run through motor systems in the brain that (as we will see in subsequent chapters) aid adaptation and social interactions and underlie our participation in sports.

Diverse information molecules underlie the organization of action; they include peptides, steroids, and neurotransmitters, and they reach back early into the origins of species, from insects to primates. Dopamine, for instance, which is essential in movement, is one fundamental neurotransmitter tied to the organization of action and thought, expectations and adjustments to changing events (picture a basketball game).

Importantly, neurotransmitters such as dopamine are tied to the organization of action, salience, cognition, and, perhaps, the prediction of events. Diverse regions of the brain (for example, the amygdala) and neuropeptides (oxytocin) are linked to social contact and play-related behaviors, and these same regions are crucial to a broad-based assessment of social context and meaning.

Sport realizes our biological capabilities through cultural expression. Action patterns are aligned with experience, neural systems are designed for the organization of action, and our language is rich with action-related events. Indeed, cephalic capability from the ordinary to the elite is *anticipatory*.

3

EVOLUTION, PLAY, AND SPORT

Although we have traced something of the history of sport in early modern human culture, no one really knows when the first true sport appeared. Which early hominin first instigated a game of catch? Did Neanderthals take turns seeing who could throw a spear the furthest? All mammals play, especially young mammals, and the spirit of play is certainly essential to sport. But play is not sport. Sport requires a set of rules that determines clear winners and losers. Play is a looser category, albeit one that may develop into sport.

We may not know when sport first became part of the human repertoire, but we can examine the evolutionary process and look at some of the human adaptations essential to modern sport.

In this chapter (and indeed, throughout this book), a recurring theme is social contact, the origins of play and sport in our evolution, and the expansion of neural function through diverse information molecules that facilitate social contact (for example, oxytocin). Sport, as an expression of our physical and mental capabilities, is a window into our biological and social evolution and their continuity in diverse human experiences.

SOCIALITY AND THE PRELUDE FOR SPORT

Several premodern hominins seem to have had some elements of what we consider culture. *Homo erectus* probably discovered and passed on knowledge about the control of fire. By about four hundred thousand years ago, *Homo heidelbergensis* seems to have carved out hut shelters

(Tattersall 1993). Sport was part of that cultural evolution. For example, bodily excellence and capabilities, evolutionary endurance, and the utilization of expanded brain size are expressed in boxing, which dates easily to antiquity in Mesopotamia.

Managing core relationships in sport is important: it takes work to keep a team cohesive. For instance, humor is a nontrivial feature. It serves many functions, including group cohesion, intimidation, and relief (consider "trash talk" on the basketball court). Laughter solidifies the group, and humor facilitates laughter (Dunbar 2010). Laughter is a key feature of our cephalic state; doing without laughter is a privation. We are social animals, and diverse regions of the brain focus on ways to facilitate social contact or at least were recruited during our evolutionary history toward this contact (Shultz and Dunbar 2006).

Brain size and expansion are tied to ecological expansion and social contact in mammals and primates (Harvey, Martin, and Clutton-Brock 1987; Jerison 1979). Brain size is linked to success in the exploration of novel environments (Sol et al. 2008); the greater the social comfort and contact, the greater the survival capability. Indeed, in social primates, social alliances greatly increase survival capability (Dunbar 2010). Social learning is fundamental because it is essential for getting a foothold in the social world of contact and survival, for alliances and pedagogy, and for finding out what to approach and what to avoid (Galef and Laland 2005). Through the sharing of resources, social order and inhibition are developed (Wrangham 1987).

Greater survivability in primates is linked to social alliances, and a reduction in cortisol is facilitated by social alliances and grooming, reducing social insecurity and building up alliances (Seyfarth and Cheyney 1984). Such alliances are a core feature of team sports and team effort. While cortisol, an adrenal steroid hormone, is vital in all activities— elevated cortisol reflects energy regulation that requires effort or attention (McEwen 1998)—extensive or extremely elevated cortisol is also an anxiety marker that social attention can alleviate.

Our evolutionary success is knotted with social-cooperative behaviors for diet and food resources, pedagogy, brain expansion, cooperative parental investments, and diverse forms of social cohesion and social concern (Kaplan, Hooper, and Gurven 2009). Sociality and object knowledge are linked. The development of language—a symbolic representation of the world in which abstract sounds stand for objects, the relations between objects, and the activities in which objects play a part—may have inaugurated the beginning of our desire to categorize. From hunter-gatherer to agrarian to participant in industrial times, humans have tried to lend structure to what they experience and to preserve it in some way (Foley 1996).

Dwelling together led to the sharing of workloads and efficient task completion, which afforded individuals more discretionary time. With this extra time, humans at some point began creating visual representations of natural objects in the form of drawings and reliefs, some of the early examples of which have survived in the deep caves of southwestern Europe.

Early humans developed many forms of instrumental expression and tool use. To go with these, we engendered numerous cognitive adaptations, some broad and some specific. Cognitive expansion and fluidity or flexibility, in which narrow adaptive abilities expand in use across diverse problematic contexts, have become a signal feature of the human mind (Mithen 1996, Lieberman 2013). In table 3.1, the relationship of cognitive development and the resulting play in childhood is outlined. Play is a cardinal feature in this evolution.

Language, with its rich syntactical elegance and expansive capacities, is a core feature of our species. Once syntactical language use emerged, our cognitive abilities seem to have increased in great measure, particularly our social discourse. This social contact and preadaptation, in turn, is vital for the formation of basic regulative events that traverse a wide range of reward within the behavioral biology of our central nervous system.

We are social animals, and many sports reflect this, but we are also solitary. The solitary walker that Rousseau (1782) wrote so elegantly about is reflected in us and in our sports and indeed is a reflection of us as whole individuals.

Two very different human motivations underlie all other fundamental human activities and experiences: the social contact essential for our survival and our sense of ourselves as individuals who are fundamentally alone (Whitehead 1919, 1929, 1938; Jaspers 1913). Sport is about teamwork and social cohesion, but it is also about solitary effort and practice and

TABLE 3.1 **Common themes in our cognitive development that underlie the origins of sport**

Infancy	1. Following attention and behavior of others: social referencing, attention following, imitation of acts on objects 2. Directing attention and behavior of others: imperative gestures, declarative gestures 3. Symbolic play with objects: playing with "intentionality" of object, playing with others, peekaboo, hide-and-seek, track and find
Early childhood: language	1. Linguistic symbols and predication: intersubjective representations 2. Event categories: events and participants in one schema; rule following, learning, sport rules 3. Narratives: series of interrelated events with some constant participants
Childhood: multiple perspectives and representational redescriptions	1. Theory of mind: seeing situation both as it is and as others believe it to be 2. Concrete operations: seeing events or objects in two ways simultaneously; sport expansion 3. Representational redescription: seeing own behavior/cognition from "outside" perspective

Source: Adapted from Tomasello et al. (1993).

more practice, and struggle and perseverance and more struggle, much of it propelled by individual desire.

For social sports, the perception of others within action is a continuous function, but this is also true in more solitary sports. We are with others, but the sense of the individual pervades in the perseverance to continue, which is knotted to ecological contexts of meaning and adaptation (Gibson and Ingold 1993; Barret, Henzi, and Rendall 2007; Mead 1934); an embodied sensibility or meaning (Varela, Thompson, and Rosch 1991; Clark 1998) pervades the sense of sport.

The mind is tied to adaptation and exploration via transactions with others, on the one hand (Dewey 1938), and expansion of capabilities into new domains of discourse, on the other. Sport is a way in which we do just that: we expand our capabilities. It is social even if the sport is individual because it is often played with others watching and reveals key motivations: motivation to win, to succeed, to achieve excellence, and to have fun. A fusion of motivational lures in the context of the attainment of form and function predominates in the expression of sport.

Sport serves a fundamental social allure, like music and art: we form bonds. Sports are primal ways to run the gamut of human expression from the brutal (boxing) to the elegant (gymnastics' high bar). All sports are about control and expansion. Cephalic expansion (Merleau-Ponty 1942, Noe 2004) is inherent throughout: probing, predicting, tuning in, expanding, and extending. In the jargon of philosophers, the events reflect "enactive realism" (Noe 2004). The sense of geometrical conception predominates, as do time, the statistical prediction of events, and expanding horizons, as we probe events with diverse cephalic capabilities or abilities.

PLAY

Imagination is a key feature of our cognitive arsenal. As a kid, I often imagined myself as Mickey Mantle. Imagination underlies sport: we need to be able to visualize ourselves performing like our heroes: this is part of the process of getting better at a skill, however much talent we may have (or lack). Such imaginary actions inhere in the mind during development, and if we are lucky they inhere through a lifetime of activity. When anchored to action, such imagination can facilitate performance.

Play, the observation and prediction of events, tagging what is familiar and what is not, is operative at almost the onset of development (Carey 1985; Keil 1989, 1979). Young infants may or may not be able to track events, evaluate events, have expectations that concur, have predictive capabilities, and impose diverse forms of categorical and continuous functions. Coherence, after all, is fundamental for survival (Tomasselo 1999).

Play facilitates social contact in all social mammals (all of whom exhibit some level of playfulness, especially as juveniles) and particularly in us. In our species, diverse events facilitate play. Diverse forms of priming is one example of facilitating social play in young children (Kirschner and

Tomasselo 2009, 2010). Underlying such developmental events are the same diverse neural generators and timing required in batting, golf, and other sports.

Our cephalic systems are endlessly anticipatory, cohabiting space in an expansive way (James 1887, Merleau-Ponty 1962). Sensory features are linked up in the brain across sensorimotor systems that reflect action and perception (Prinz and Barsalou 2000, Martin 2007, 2015). Scanning the terrain for relevance and developing behavioral inhibition for greater behavioral regulation are both essential in sport.

Our astonishingly fast cultural evolution over the last ten thousand years—which includes the taming of fire, the development of agriculture, and the onset of technological innovation—provides the context of play, exploration, and the beginning of sport. During our development, we go easily from play to sport. Play is by definition freer and less rule bound than sport, but play is not without competition (think of two puppies wrestling; they are clearly having fun, but they are also establishing dominance boundaries). By age three or four, children will follow simple rules in a board game such as Candy Land and by five become passionate about rules and rule breaking, even in play (Caillois 1939, Kagan 1984).

In play, we endlessly explore, which is a key feature of our evolutionary success. While sensorimotor systems and coordination are not time locked to all cognitive capabilities inherent in exploration, putting objects in one's mouth as an early means of exploring one's environment demonstrates that relevance begins early (Piaget 1954; Keil 1979, 1989). There are disputes about when in ontogeny this occurs, but exploration, play, and discovery are locked into a nexus of developmental trajectories where tracking one's actions and surroundings is a primary feature. What you are looking at matters in getting a foothold in your surroundings. There is a transition from play to practice and the progression from play to sport.

Deliberate play is done for its own sake, is enjoyable, can occur in a variety of settings, and does not require rules or adult involvement, whereas deliberate practice is done to achieve a future goal, is not necessarily enjoyable, often occurs in specialized facilities, and needs explicit rules and often adult involvement. Play and sociality are preludes for sport, and practice and more practice is the underlying theme in sport. And sport experiences tend to be influenced by one's early training and

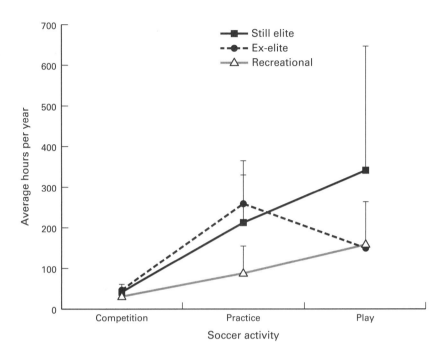

FIGURE 3.1

Mean (and SD) average hours per year in soccer activity in six- and twelve-year-old soccer players (still-elite, ex-elite, and recreational groups).

Source: Adapted from Ford et al. (2009).

experiences. Play has been linked to brain size but not consistently (Iwaniuk, Nelson, and Pellis 2001; Bickart et al. 2011).

The continuous theme is play and practice and the expression of motor/cognitive capabilities in neural function (Graybiel 1998). We track others, and we track what they are involved with. We tag what others pay attention to and what their beliefs and desires are, and we form alliances with those we feel share traits similar to ours; this is no doubt key to our evolutionary success. Sports evolved in our cultural ambiance of going from play to games to sports and to embracing form and function within set rules.

Predator play in the development of form is common across mammals and absent in, for instance, fish. Play matters for learning and for social learning in particular. A number of adult predators are noted for, under some conditions, playing with their prey before they kill it; killer whales with seals and housecats with mice, for example. Whether this is truly play is another issue, but it certainly looks like it: it is not necessary to the act of predation, and the predators appear simply to enjoy it. In some cases this play may have a role in teaching the young, but in many observed instances no young are present.

Play is such a vast part of the human condition (Caillois 1939) that we would not understand ourselves without it. But we are not alone (Bekoff and Byers 1998). Indeed, play behavior is phylogenetically fairly ancient; play is noted, for instance, in perhaps birds (for example, common ravens [Heinrich and Smolker 1998]). Play behaviors are rampant in mammals (Fagen 1974, 1981; Beach 1979).

Play behaviors are particularly clearly evident in social animals, and play behavior may last a lifetime. Play behaviors reflect diverse changes in neural capability and function (Siviy 2010) that are tightly linked to developmental trajectories and life histories (Bekoff 1995; Siviy and Panksepp 2011; Vanderschuren, Niesink, and Van Ree 1997). Play is tied to the development of skill formation, physical and mental capabilities, and social contact. Cognitive/motor expression is strengthened in play behaviors, as are changes in neural function (Byers and Walker 1995, Siviy 2010).

Nature is rich in examples of animal play behavior. Play is a common event: the aesthetics of dolphin leaps and social-cooperative behaviors, the dogs at city parks, the greetings of animals. But play turns easily into aggression in many forms of play in both juvenile and adult animals. Play behaviors in animals is tied to vocalization patterns (Knutson et al. 1998). Among mammals, the larger the brain, the more playful the mammals are, particularly with sex play (Iwaniuk, Nelson, and Pellis 2001).

Diverse information molecules (endogenous opioids, cortisol, etc.) are elevated during play behaviors, as has been seen in studies with rats, and such molecules are no doubt present in mammals in which there is a gestational period of play behaviors. Peptides such as oxytocin and the opioids are linked to diverse forms of play behavior (Nelson and Panksepp 1996). The presence of these molecules is particularly evident in rough-and-tumble play behaviors.

Play behaviors also are knotted to social contact, and facilitating these social relationships are oxytocin, opioids, and vasopressin, which are expressed in diverse regions of the brain. Within the developmental trajectory of play, imaginary others, peekaboo, and problem solving through practice in sport, all forms of human activity and development are enlivened. Problem-solving capabilities blend into our cultural footing with others.

Regions of the amygdala appear to be essential for social attachment and avoidance. For instance, one can view the enlargement of the lateral amygdala in humans, which is closely tied to neocortical function, as an evolutionary change favoring sociality (LeDoux 1995, 2016; Swanson 2000; Amaral et al. 1992; Aggleton 1992, 2000; Amaral et al. 2002; Emery 2000). The largest nuclear region is the basal lateral region (Stephan, Frahm, and Baron 1987; Schumann and Amaral 2005).

The greater the social complexity of the species, the greater the amygdala volume. The amygdala volume of humans is almost 0.6 percent; those of chimpanzees, bonobos, and gorillas are all about 0.1 percent. Interestingly, the amygdala, in addition to hypothalamus size, is thought to be linked to social play in primates. Lewis and Barton (2006), among others, have suggested that social play (and perhaps sport, for that matter) is tied to this expansion; affective social contact is, after all, a feature of games and sport as well as of the socialization process generally.

Primates who spend a low percentage of their time budget on social play tend to have smaller brains (specifically, smaller amygdalas and hypothalami) than those who spend more of their time playing. For instance, the common marmoset, which only spends 1.32 percent of its time budget on social play, has a mean adult brain size of 7.9 grams and an amygdala size of 113 cubic millimeters; the western gorilla, which spends nearly 15 percent of its time playing, has a mean adult brain size of 505.9 grams and an amygdala size of 2,754 cubic millimeters.

Core cephalic capabilities and events include the exploration of diverse terrains, the development of language and other cognitive communicative competences, omnivorous feeding and predatory patterns, tool use, bipedal structure, and social abilities. All these aspects of evolution figured importantly in our development and colonization of the globe from our origins in Africa (Leakey and Lewin 1977; Mellars 2006a, 2006b; Foley 2001).

OXYTOCIN, SOCIAL ATTACHMENT, AND SPORT

Oxytocin is an information molecule that, depending on the circumstances, can favor group cohesion or elicit aggression toward others. As a neuropeptide, oxytocin facilitates social attachment, especially between relatives and members of the same group. Oxytocin tends to elicit trust behaviors generally, but familiar faces elicit greater approach behaviors. The team, the group, underlies social contact, and neuropeptides such as oxytocin facilitate such relationships; this would thus also underlie the contact between individuals on sports teams. However, oxytocin also underlies antagonism toward those not in the local niche. When it comes to those outside the familiar social group, in animal studies oxytocin has been seen to elicit aggression, especially territorial aggression. When viewed from the perspective of competitive sport, oxytocin may be responsible for the aggression often observed between members of opposing teams (Carter 2014).

Social contact and the creation of alliances in sport are necessary and are seen in all competitive sports. Social cohesion between fans of the same team and between team members, as well as aggression toward those not on the team, are common sports currency. Belonging to part of a group is a nontrivial aspect of the human condition that underlies sport. Social longing is primordial and integral to survival.

A balance of oxytocin and vasopressin (Neumann and Landgraf 2012) underlie many behaviors, including social ones. Oxytocin, a phylogenetically ancient amino acid, is expressed in diverse end organ systems as well as the brain, and there are variants of oxytocin receptors in both the brain and other organs. It is regulated in the brain by steroids (Choleris et al. 2007). It is linked to vasopressin in its origins and has fundamental links to fluid balance. Along with vasopressin, it has a rich homological history. But it also figures in play, and play is a precursor of sport.

The medial region of the amygdala is one such area of the brain linked to social perception and to gathering the gist of events quickly. This fast and frugal heuristic response, which has worked effectively for millennia (Adolphs 1999, 2001; Gigerenzer 2000), is crucial in competitive sports.

TABLE 3.2

Bovine -168	C A T A A C C T **T G A** C C
Sheep -161	C A T A A C C T **T G A** C C
Mouse -174	G A T **G A C C** T T **G A** C C
Rat -168	G G T **G A C C** T T **G A** C C
Human -164	G G T **G A C C** T T **G A** C C

Details of the approximately 2160-bp region (composite hormone response element) of the upstream OT gene promoter conserved across five species, including the sequences of the response elements estrogen response element (ERE), chicken ovalbumin upstream promoter transcription factor I (COUP-TF), and steroidogenic factor-1 (SF-1).

Source: Gimpl and Fahreholz (2001).

The amygdala is linked to social perception and is innervated by olfactory information.

Genomic manipulations of specific estrogen receptors and their interaction with the amygdala affect social behaviors. Tests where oxytocin expression was knocked out showed that subjects did not habituate as readily to familiar conspecifics without the oxytocin gene (Ferguson et al. 2001, Choleris et al. 2006).

Importantly, infusions of oxytocin into the brain activate enkephalin receptors (one form of endogenous opioid receptor tied to social reward), and the nucleus accumbens, a brain region linked to motivation, is tied to social contact and social avoidance (Kelley 1999; Peciña, Smith, and Berridge 2006; Lim, Bielsky, and Young 2005). Events such as these are likely inherent in social sports activity.

The exciting thing about oxytocin is the varied role that it appears to play in many features of social attachment: social grooming, looking at others, eye contact, shared cooperative behaviors, trust, approach behaviors, and bonding (Insel 2010). Oxytocin is just one of the information molecules in the brain that underlie the social-cooperative behavior that pervades social sports (Zak, Kurzban, and Matzner 2005). Indeed, oxytocin enhances the responses to diverse forms of facial expression, sounds, and social contexts that underlie sports.

HUMAN EVOLUTION

About eight million years ago, the first hominids made an appearance (Coppens 1994). They were small in stature but had comparatively large brains—about 30 percent larger than a chimpanzee brain (Holloway 1980). This hominid, commonly called *Australopithecus*, stood upright and is believed to have been both bipedal and arboreal.

We have since accommodated ourselves to diverse situations. We have adapted to live in cold climates; learned to explore; developed tools, clothing,

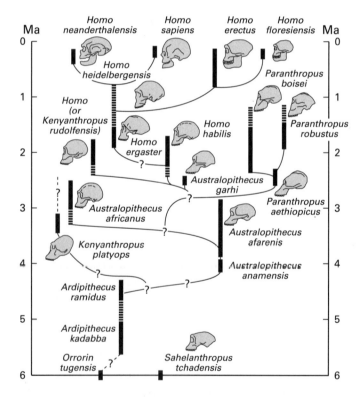

FIGURE 3.2

One depiction of hominoid evolution.

Source: Adapted from McHenry (2009).

and houses; tamed fire; and domesticated animals. These are all features of cultural evolution. But first, let us look at our biological evolution.

The gaps between different hominins are as striking as the similarities. Their evolutionary trajectory is not one dimensional or progressive but highly diverse, perhaps with bouts of stability or stasis within "punctuated change" (Gould and Eldridge 1977).

Before hominins even made an appearance, primate diversification was marked. Some primates cohabited the earth at the same time and dispersed, although most did not survive; selection for capabilities and favorable contributions made all the difference (Darwin 1859, 1871).

Multiple hominins also appear to have competed at the same time. What was selected for was not the brain per se but some combination of features out of which an evolving brain took precedence (Foley 1995, 1996).

The progression is not a continuous function; there are starts, stops, variations, and, of course, extinctions. All of the above, with the exception of *Homo sapiens*, became extinct, as likely did many others we are not aware of.

Characters unique to all aspects of *Homo*, as distinct from other hominids:

1. Increased cranial vault thickness
2. Reduced postorbital constriction
3. Increased contribution of occipital bone to cranial sagittal arc length
4. Increased cranial vault height
5. More anterior foramen magnum
6. Reduced lower facial prognathism
7. Narrower tooth crowns, especially mandibular premolars
8. Shorter molar tooth row

(Adapted from Wood 1992)

Many differences are obviously at a molar level across species. What is different at a molar level of analysis, for instance on the morphological side of *Homo*, are changes in cortical expression, which, unlike changes in facial and dental morphology, leave no fossilized traces (Wood 1992; Mellars 2006a, 2006b; Holliday 2002).

CEPHALIC EXPANSION

Cephalic expansion set the stage not only for our visual acuity but also eventually for technological creations, expanding our sensory systems. Seeing by magnifying became an evolving theme as our capacities were extended, and we turned from managing nature toward understanding it.

Our evolution as a species depended largely on forming stable social groups. This reliance on social groups continues to play a critical role, including in sports, for both the athletes and the spectators. Thus the focus on the group, the team, is an important adaptive expenditure (Dunbar 2010).

We do know something about the difference in cortical mass in diverse hominins. The trend, not surprisingly, is toward greater weight across the evolution of *Homo*. Table 3.3 depicts suggestions about cranial capacity, encephalization, and brain size in extinct hominins and humans (Foley 2006).

Dramatic changes in climate during the late Pliocene (Mellars 2006a, 2006b) may have driven our biological evolution to include an expansion

TABLE 3.3 The encephalization quotient (EQ) of various extant and fossil species

SPECIES	DATE RANGE	EQ*
Pan troglodytes	Extant	2.00
Australopithecus afarensis	3.9–3.0 M**	2.50
Australopithecus africanus	3.0–2.4 M	2.70
Paranthropus boisei	2.3–1.4 M	2.70
Paranthropus robustus	1.9–1.4 M	3.00
Homo habilis	1.9–1.6 M	3.60
Homo ergaster	1.9–1.7 M	3.30
Homo erectus	1.8 M–200 K***	3.61
Homo heidelbergensis	700–250 K	5.26
Homo neanderthalensis	250–30 K	5.50
Homo sapiens	100 K–present	5.80

* EQ expresses brain size relative to body mass and thus is a measure of relative brain size.

** million years ago

*** thousand years ago

Source: Adapted from Power and Schulkin (2009).

of our cephalic capabilities. Temperature, landscape, and prey variability may have led to an increased reliance on cephalic capabilities that would allow for the exploration of unfamiliar terrain, the tracking of prey, the finding of seasonal food sources across great distances, and the formation of a community to share in the responsibilities required for survival. Small-group contact was essential to our immediate survival and eventual evolution. While cephalic expansion undoubtedly played a role in this, cephalic evolution is not continuous; small changes interspersed with longer-term stasis may have been evolutionary currency (Gould et al. 1996, 1999, 2001). Cephalic expansion is tied to social complexity as well as to longer infant gestation and juvenile developmental periods. This is the context for learning from others, play pedagogy, exploration, and, eventually, sport.

TOOLS AND TOOL USE

Without tools there would be little sport. We may have begun by playing with sticks and rocks. The expansion of cephalic function provided physiological and behavioral regulation. An expanded motor system with diverse cognitive capacities was no doubt pivotal in our evolutionary ascent. It is not just that the evolution of the cortex or brain is tied to social function; tool use and other diverse abilities are also knotted to the fact that we are social animals. Thus, innovative tool use has been linked to an expanded brain ratio, and tool use figures in almost all sports.

Tool use and its elaboration are reflections of expanding motor capacity, and any tool comes rich with cognitive possibilities. Of course, we are not the only species to use tools. Birds use sticks to build nests, chimpanzees and other primates strip twigs of their leaves and use them to fish tasty termites out of their mounds, and some chimps on occasion use tree branches as spears to kill prey.

Significantly, the construction of tools by primates such as chimpanzees is often tied to a social context. In their tool use, we can see the first glimmer of our species' ability to use objects as tools to facilitate the expansion of our social milieu, for it is the expansion of cephalic function that underlies the tool use that serves physiological and behavioral regulation. The size of the cortex across our species is tied to a great array of capabilities: spatial and mathematical capabilities, language and auditory capabilities, and play and sport behaviors.

Tool construction (including tools used for sports) and cognitive/motor systems are both essential to the evolution of the brain. Tool use is an expression of an expanding cortical motor system in which cognitive systems are endemic to motor systems. For instance, regions of the brain are prepared to recognize differences between different kinds of objects and their uses, of which mechanical tools are an important subset. And of course, tools are used in sports.

Furthermore, frontal motor regions have been linked to the motor features of tool use. Importantly, many species use diverse tools in adapting to their environment, which reflects the evolution of cortical and subcortical systems in the brain that participate in tool use, tool making, and tool recognition (Johnson-Frey 2003, Martin 2007, 2015).

The cultural evolution of hammers and cutting implements and spears is later paralleled in the development of sticks, bats, golf clubs, and tennis rackets. The strict bifurcation between biological and cultural evolution is misleading and false, and the modern rise of epigenetics further undermines this distinction. Cultural expression in the construction of objects, tools, and languages are features of our evolution (Dewey 1925). We keep expanding our capabilities, and this is reflected in sport. Sport as a cultural expression is as natural as the leaf outside my window.

Most sporting tools have evolved significantly. As cultural evolution merges with cephalic and bodily capability, what we see across the range of human experience is the decrease of metabolic costs in human mobility, driven by our tool making and using. Skiing, a surprisingly early development, is one example (Formenti, Ardigó, and Minetti 2005). Sport is both an example of our ingenuity with tools and an instance of our ability to expand our capacities evolutionarily.

CONCLUSION

Brain growth is not just a human feature but a common currency across evolution, in one way or another. Flies have no cortex, but one marvels at what evolution has allowed them to do—invertebrates are so utterly different from vertebrates in how they have evolved. In speaking of verte-brates, however, the expansion of the forebrain is extraordinary.

The cortex is expanded in mammals, and the basal ganglia is compara-tively large in birds. Some marsupials lack the massive connective tissue (the corpus callosum) that binds the two hemispheres much more closely than do the anterior and posterior commissural connections. Consider how our understanding of brain structure has evolved over time and how our understanding of brain function has developed with regard to social milieu and anticipatory regulation.

Our species devotes vast cognitive/neural capital toward the social milieu (Gazzaniga 1985, 2005) and to an endless array of anticipations. Concerns about where the ball will arrive in soccer, whether the shot will go into the basket, and where the defense will be positioned are all manifestations of this generic anticipation-and-control feature of the brain (Yarrow, Brown, and Krakauer 2009). Sports evolved to facilitate social contact and social solidarity, and it thereby facilitates biological capabilities and social expression.

4

GENETICS, EPIGENETICS, AND TALENT

I was recently out with my daughter, the other swimmers on her team, and their parents. In most gatherings of parents you wouldn't expect to find sports stars. But two of the parents in this group had been professional athletes, and most of the rest had been successful amateurs. Clearly, both sense and culture have something to do with this. To succeed at anything, you have to have the essential capability, and you have to have had cultural access to the tools needed for success. Genes underlie all human activity; they build the peptides and steroids and neurotransmitters required for the initiation of action.

Elite athletes are geniuses (Zimmer 2005) possessing elegance, excellence, and the mindset for winning: that ability to want to win *and* to be able to capitalize on their advantages. You often hear in sport that so-and-so does not have the mindset to make things work, to persevere when it counts, to seize an opportunity, to sustain momentum when things are bad, etc. All those characteristics are part of the mindset of "being a winner." Few reach that ideal. But we know it and can point to it when we see it. Genius in sport expresses that mindset.

In this chapter we look at genes and their expression in athletic capability. There is no one athletic gene; there is a general confluence of specific and general capabilities that converge on athletic expression. Such events reflect experience, culture, and epigenetic expression and exemplify the absolute continuity of biology and culture: each a reflection of the another.

GENES AND HUMANS

Genetics and evolution are crucial themes in biology and for understanding ourselves. The Austrian friar Gregor Mendel pioneered what became genetics by studying peas, and over time we have come to understand that DNA is the common currency in inheritance (Watson and Crick 1953).

Simply put, DNA codes for gene products are transcribed by RNA. We look at the production of the material (information molecules, receptors, etc.) to understand what makes them tick. Genes help make many products. There are between twenty and twenty-five thousand genes in humans. This number is expanded considerably by post-translation events. Diverse peptides, for instance, have a number of iterations in organ systems within and across species. Indeed, gene regulation of such products is commonplace throughout the life process in many species, including plants.

In this age of molecular biology, we are capturing the genome of most species, including ourselves, and in particular the genes that underlie our cephalic development. Table 4.1 shows features that make us different from other primates.

GENES AND TALENT

Genius and capability are tied to genes and circumstance, temperament and opportunity. Very few of us can match Tom Brady, Serena and Venus Williams, Joe Montana, Michelle Wie, or Tiger Woods. Their raw ability, the astonishing focus they bring to bear to turn that ability into excellence, and the sheer ingenuity of their anticipatory skills are dazzling (Babiloni et al. 2010).

This exceptional raw capacity may largely be a matter of genetic luck. Several twin studies have explored the factor of genetic luck, and the studies concluded that there was a genetic link to physical capability or athleticism. Mustelin et al. (2012) demonstrates this link by finding that heritability for sports activity is 64 percent. But capability has to be coupled with rigorous training and a supportive social context, for example, a family that encourages and sustains: as the cliché goes, the way to get to Carnegie Hall is to practice, practice, practice. We all know that talent

TABLE 4.1 Distinguishing features of humans

PHENOTYPIC FEATURE	HUMAN-LINEAGE-SPECIFIC TRAIT	POSSIBLE EVOLUTIONARY ADVANTAGES
Brain growth trajectory	Prolonged postnatal brain growth and delayed myelinization period; enhanced cognition	Allowed creation of novel solutions to survival threats; increased the critical period for learning new skills; facilitated emergence of uniquely human cognitive skills
Brain size	Increased brain–body size ratio; enhanced cognition	Allowed creation of novel solutions to survival threats; improved social cognition
Larynx location	Portion of tongue resides in throat at level of pharynx; larynx descended into throat	Helped in the development of spoken language
Eccrine sweat gland density	Higher density of eccrine glands; enhanced sweating capacity	Enhanced cooling ability; allowed protection of heat-sensitive tissues (such as the brain) against thermal stress; facilitated endurance running
Capacity for endurance running	Improved energy use during periods of high energy demand; increased capacity to transfer energy (in the form of glycerol) from fat stores to muscle; anatomical changes relating to running ability	Allowed persistence hunting to emerge for accessing the benefits of increased meat consumption; increased range of food sources; the resulting improved diet may have facilitated brain evolution
Labor	Earlier onset and longer duration of labor	Protected the child and mother from damage caused by increased head circumference
Lacrimation	Emotional lacrimation (crying)	Enhanced emotional communication within social groups; increased affective communication
T-cell function	Relative T-cell hyper-reactivity	Enhanced immune function
Thumb	Increased length; more distally placed; larger associated muscles	Allowed creation of more detailed tools; allowed manipulation of objects on a finer scale

Human-lineage-specific (HLS) traits are phenotypes of the human lineage that arose after the split from the Pan lineage. A substantial number of forces are likely to have contributed to the development and maintenance of these traits; several examples are listed here. Plausible forces commonly discussed are macro- and microlevel climate changes that occurred frequently over the course of human evolution and that may have selected for rapid HLS changes to survive novel climatic challenges.

Source: Adapted from O'Bleness et al. (2012).

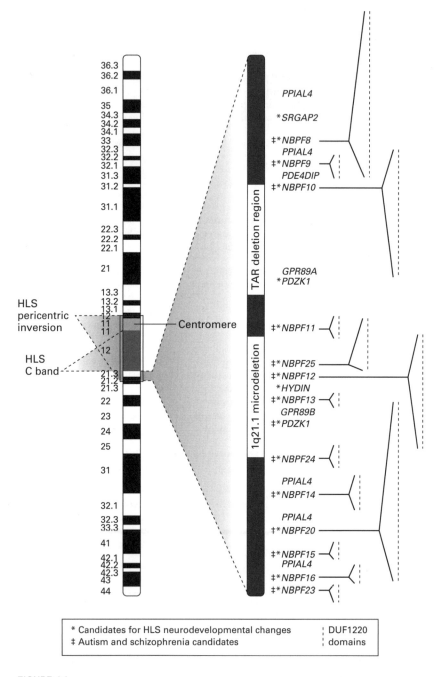

FIGURE 4.1

Evolution of genetic and genomic features unique to the human lineage.
Large cytogenetically visible changes in the genome structure were among the first
human-specific genomic changes noted between humans and great apes. More recently,

goes only so far, and that's not actually very far at all once one gets to the major leagues. The world is littered with talent that has never been fulfilled. Discipline, perseverance, support, opportunity, and what used to be called character are needed to fulfill the promise of talent.

I don't believe there is a single sport gene or even several. But what there is is a confluence of capabilities sculptured on a context and a sport (O'Bleness et al. 2012; Lippi, Longo, and Maffulli 2010; Ehlert, Simpson, and Moser 2013). Like art, you know it when you see it, and it is not just a score in a record book (although that matters).

Genius in anything, as we all know, is rare, but lots of capabilities are inherent. Socrates in the *Meno* argued that geometrical capability is inherent in both the educated and less educated, but Euclid or Descartes enjoyed strokes of genius with regard to geometry. Sport, however, requires a myriad of capabilities that cohere in a suitable environment, with the measured meaning of form and content.

Sport also requires a balance of thought in action. In experiments with golfers, experts given longer amounts of time to visualize their putts did worse on the actual putting than those given less imagining time. Novices, however, noticeably did better when given more visualization time. Apparently, once a skill has been honed, it is possible to "overthink" the process—it's better to let muscle memory take over. "Stop thinking!" is a common coaching call across all sports; athletes need to let muscle habit, or what C. S. Peirce called "frozen habit" and what we might call codified cephalic capabilities in muscular tissue, take hold and take precedence.

---

it has been determined that these regions harbor more importance than just being human-specific heterochromatin. Indeed, such regions are frequently adjacent to regions that are greatly enriched for evolutionarily recent gene duplications and that often function as gene nurseries. For example, the 1q21.1 region of the genome, which lies adjacent to the human-specific 1q12 C band and within the human-lineage-specific (HLS) chromosome 1 pericentric inversion, has undergone substantial genomic enlargement owing to numerous HLS copy number expansions within the region, as shown by the green bands in the figure. Numerous findings have identified the 1q21.1 region as being highly enriched for HLS copy number expansion, including striking HLS copy number increases of DUF1220 protein domains.

Source: Adapted from O'Bleness et al. (2012).

TABLE 4.2 **Partial list of genes and genetic elements showing human-lineage-specific changes**

GENE OR ELEMENT	MECHANISM OF CHANGE	PROPOSED PHENOTYPE	PHENOTYPIC CERTAINTY
Androgen Receptor (*AR*)	Deletion of regulatory DNA	Loss of sensory vibrissae and penile spines	Likely
Apolipoprotein C1 (*APOC1*)	Pseudogene	Unknown	Not applicable
Aquaporin 7 (*AQP7*)	Copy number increase	Energy use	Plausible
Asp (abnormal spindle) homologue, microcephaly associated (*ASPM*)	Positive selection	Increased brain size	Plausible
CDK5 regulatory subunit associated protein 2 (*CDK5RAP2*)	Positive selection	Increased brain size	Plausible
Cholinergic receptor, muscarinic 3 (*CHRM3*)	Novel exon	Change in human reproduction	Plausible
Dopamine receptor D5 (*DRD5*)	Copy number increase	Regulation of memory; attention; movement	Likely
Forkhead box P2 (*FOXP2*)	Amino acid change	Speech and language development	Definite
Human accelerated conserved noncoding region 1 (*HACNS1*)	Positive selection	Changes in anterior wrist and thumb	Likely
Myosin heavy chain 16 pseudogene (*MYH16*)	Pseudogene	Craniofacial musculature	Plausible

Genes that have associated human-lineage-specific (HLS) traits are listed with an assigned level of certainty with regard to their impact on human uniqueness. Certainty ranges from plausible (that is, the association is still hypothetical on the basis of what is known about the gene) to likely (that is, there may be a disease association or animal-model evidence to substantiate the claim) or definite (that is, there are multiple lines of supporting evidence). In cases in which the gene has been implicated in a disease but an HLS phenotype has not been proposed, the certainty column is not applicable. In addition, associated disease links are given if known.

Source: Adapted from O'Bleness et al. (2012).

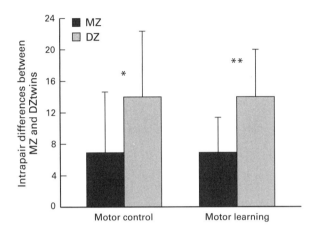

FIGURE 4.2

Mean and standard deviation of intrapair differences between monozygotic (MZ) and dizigotic (DZ) twins in motor control and motor learning.

Asterisks indicate significant differences (paired t test; *P = 0.05 and **P = 0.01, respectively).

Source: Adapted from Missitizi et al. (2013).

A great deal of genius in sport, therefore, is hitting the right balance. Think of the genius of the punter, who has to assemble the right balancing pressure and accuracy but who must also put aside the paralyzing knowledge that points rest directly and solely on his shoulders—one kick can decide the game. Not overthinking is an expression of genius in sport.

Diverse neural circuits are tied to memory; after all, cultural evolution and our sense of being in the world are tied to what objects represent and where they are (Ungerleider and Mishkin 1982). One view of consciousness is that it is deeply related to our short-term memory, the needs of the moment, imagined or real (Desimone 1996). Of course, in the context of anticipating what might come next, both short- and long-term memory are necessary to excel in sport.

There are two primary action systems in the brain that contribute to our knowing "what" something is (the ventral stream) and "where" it is (the dorsal stream). These systems permeate all areas of our lives, including sports. The visual-processing streams in the brain pervade our perception and action (Goodale and Humphrey 1998). From grasping a pencil to catching a football, the visual stream organizes our visual input into coherence.

The dorsal stream is responsible for recognizing and estimating the location, position, and speed of objects prior to executing the action (Goodale 2014) of the "grasp" and "use" systems (Binofski and Bauxaum 2013). The innovatively recognized "grasp" system is hypothesized to be immediate and short lasting after initiation because of the reflexive innateness that grasping exudes. However, the "use" system is gradual and slightly longer lasting because it is supported by an already established conceptual knowledge of the object that will ultimately be grasped and used for its purpose. Although the dorsal—with the codependent "grasp" and "use" systems—and ventral streams are independent, the two streams work effortlessly and simultaneously together to prepare action, physical capability, and, ultimately, sports.

FIGURE 4.3

Goalkeepers utilize various strategies and skills when practicing and playing: tracking multiple events within the central and peripheral visual fields, using information from a large visual field, and quickly processing all of this information.

Source: Faubert and Sidebottom (2012); Public Domain.

TABLE 4.3 Major candidate genes associated with human athletic performances

Endurance Capacity	*PPARD*
	Nuclear respiratory factors (*NRF2*)
	PGC-1 alpha
	HIF-1 alpha
	EPAS-1 and *HIF-2 alpha*
	Hemoglobin
	Skeletal muscle glycogen synthase (*GYS1*)
	ADRB2
	CHRM2
	VEGF
	Maximum oxygen intake (*VO2max2*)
	Proliferat-activated receptor a gen (*PPARA*)
	a-actinin-3 (*ACTN3*)
	ARG(R)577Ter(X)
	3'-phosphoadenosine-5'-phosphosulfate synthase2gene (*PAPSS2*)
	ACSL1 polymorphism (*rs6552828*)
	Anti-dorsalizing morphogenetic protein (*ADMP*)
	Histidine (*HIS*)
	Glycine (*GLY*)
	Myostatin (*GDF8*)
	Interleukin 6 (*IL6*)
	NO
Muscle Performance	*CK-MM*
	ACTN3
	MLCK
	ACE
	AMPD1
	IGF-1
Tendon Apparatus	*ABO* blood group
	COL1A1 and *COL5A1*
	TNC
Psychological Aptitude	Serotonin transporter gene (*5HTT*)
	BDNF
	UCP2

Source: Adapted from Lippi, Longo, and Maffulli (2010) and Ehlert, Simpson, and Moser (2013).

For goalies in soccer or hockey, rapid-fire information processing is the name of the game. Their visual field is expansive; they hover over space and the objects in it (Merleau-Ponty 1962). The visual field narrows in focus toward the goal object (Desimone 1996, James 1890).

In simulation tests of visual activity after training on a task on thresholds, responses in hockey and rugby players were far more expansive and impressive than in controls (Faubert and Sidebottom 2012).We understand the capabilities that a talented athlete has to possess. Our knowledge of the possible genes linked to elite athletic abilities has become a little clearer, and that list has entered our scientific lexicon.

EPIGENETICS

Epigenetics are the changes in DNA methylation given a subject's activity and life experience. Such changes are tied to both genes and the regulation of genetic changes in practice and sport. Histone modification is one cornerstone of modern epigenetic research. It is a new science, one producing more noise than results as of yet, but some of the results are intriguing. In relation to sports, researchers have examined a number of markers of excellence, focusing on endurance (Eynon et al. 2011).

Endurance in running is a key feature in our evolutionary success as a species (Lieberman and McCarthy 2007; see also chapter 6). Endurance is enhanced by the effects of exercise and lifestyle, but a number of genes contribute to human endurance abilities. Several of these are growth-factor compounds (for example, insulin), and a number of others are transmitters (for example, adrenergic receptor b3). Several genes have been linked to endurance. These genes are also tied to epigenetic changes, another of which is the angiotensin-converting enzyme; angiotensin is tied to fluid balance (Fitzsimmons 1998, Denton 1982).

Physical activity, like other forms of human action, directly affects genomic expression (Booth, Chakravarthy, and Spangenburg 2002). The hypothetical figures in figure 4.4 A and B depict changes in genetic and epigenetic expression in athletic performance (Ehlert, Simon, and

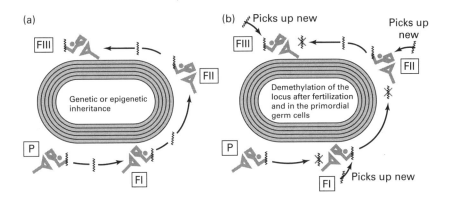

FIGURE 4.4.A

Genetic and epigenetic development over four generations: (a) Genetic or epigenetic inheritance. The schematic analogy of transgenerational genetic or epigenetic
transmission displayed as a relay run. Each runner represents one generation. The
relay baton symbolizes one specific transmitted allele or, in the case of epigenetic
inheritance, methylation status. The allele is passed on unchanged over generations,
starting with the parental generation. Of course, the probability of transmitting this
allele is only 50 percent because of the haploid germ cells. In cases of nonmendelian
epigenetic inheritance, the possibility that an inherited epigenetic mark is phenotypi-
cally expressed is further limited, although in the case of our relay-run analogy, the
runners still may transmit the baton. (b) Loss of methylation: DNA–de novo methylation displayed as a relay run. The baton, symbolizing a single DNA-methylation
status, is not passed on to the next generation but is erased after fertilization and in the
primordial germ cells of the embryo. Every generation picks a new baton, symbolizing
de novo methylation throughout ontogenesis. FI–FIII, filial (offspring) generations
one to three; P, parental generation.

Source: Adapted from Ehlert, Simon, and Moser (2013).

Moser 2013). This represents the changes in biology via culture, and such
changes can be made possible via sport.

Sport's balance of power and beauty, since the advent of ritualized
sports activities, requires mindfulness. Being mindful or attaining mindfulness is a broad, allusive term used in disciplines from meditation to
therapy to imply a focus on what counts and an exclusion of what does
not; this focus is something that elite athletes in any sport, be it martial

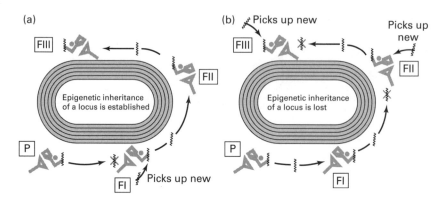

FIGURE 4.4.B

Hypothetical establishment or loss of epigenetic inheritance.

The schematic analogy of the hypothetical establishment (a) or loss (b) of transgenerational epigenetic inheritance displayed as a relay run. The baton, symbolizing methylation status, is lost from the parental to the filial-one (offspring) generation. In this generation, the methylation status could become fixed by environmental factors and is neither demethylated following fertilization nor in the primordial germ cells of the filial-two generation. FI–FIII, filial (offspring) generations one to three; P, parental generation.

Source: Adapted from Ehlert et al. (2013).

arts or snowboarding, seek to attain. Indeed, we know from studies of human information processing that enhanced attention leads to the exclusion of material that matters less, homing in on what counts, minimizing outside noise, and enhancing focus (Desimone 1996, James 1890).

Neurologically, enhanced attention is tied to the narrowing of neuronal assemblies; we can observe less neural expenditure and enhanced, focused neural activity. That is the adaptive side of neuronal assemblies: consider the focus of the archer, the pitcher, the goalie. However, being able to zoom in with unremitting focus is necessary, but still not sufficient, for sports excellence.

Epigenetics arose with the onset of the molecular integration into broader-based and regulatory biology (Dobzhansky 1962, Holliday 2002, Crews 2008). "Changes in gene expression without alteration of the underlying DNA" is a workable definition of epigenetics and is perhaps at the heart of phenotypic alterations in development contingent on context and circumstance (Dolinoy, Weidman, and Jirtle 2007).

One mechanism that may underlie epigenetic gene regulation is methylation and demethylation (that is, the silencing of or enhancement of the expression of genes) (Holliday and Ho 1998). Demethylation prevents transcriptional expression, and one result is the silencing of gene expression.

Figure 4.5 depicts a putative mechanism for silencing genes in development (Holliday and Ho 1998) and for putative chromatin regulation (Keverne and Curley 2008), including that of oxytocin (Harony-Nicholas et al. 2014). Motion or social perception has been linked to oxytocin-receptor DNA methylation (Kumsta et al. 2013; Jack, Connelly, and Morris 2012).

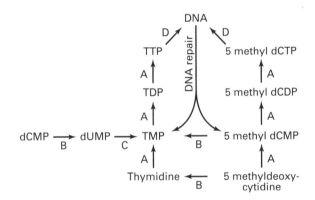

FIGURE 4.5

A mechanism for silencing genes in development and for chromatin regulation. Pathways of uptake of pyrimidine nucleotides TMP and 5-methyldeoxycytidine monophosphate (5-methyl-dCMP) into DNA. A, kinases; D, DNA polymerase; B, deaminases; C, thymidylate synthetase. Aminopterin blocks the conversion of dUMP to TMP. Cells can grow in the presence of aminopterin if thymidine (and hypoxanthine) are provided exogenously (HAT medium). 5-MethyldC can substitute for thymidine if one or both deaminases are active (HAM medium). Strains resistant to BrdU, which lack thymidine kinase, can grow in HAM medium only if 5-methyl-dCMP deaminase is active.

Source: Adapted from Holliday and Ho (1998).

Oxytocin is linked to food sharing, social bonding, and the tracking of related (versus unrelated) individuals. Cooperative behaviors such as hunting were likely stimulated by oxytocin expression. Changes in demethylation are linked to the transmission of such social behaviors to offspring in cross-fostering experiments. Indeed, manipulations of demethylation influence the transmission of this social behavior and can even be reversed later in life (Weaver et al. 2004). This is now being linked

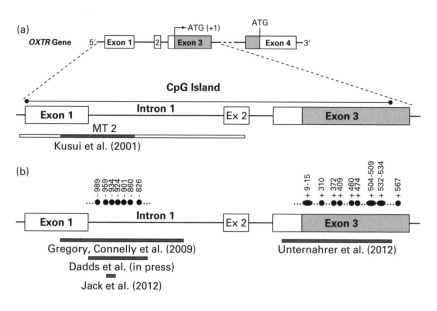

FIGURE 4.6

Panel A shows the genomic organization of the oxytocin receptor gene (OXTR). The OXTR gene is located on chromosome 3p25–3p26.2, spans 17 kb, and contains three introns and four exons, indicated by boxes. The protein-coding region is indicated in gray (ATG denotes the transcription start site, and TGA denotes the stop codon). The enlarged section at the bottom of panel A shows the location of a CpG island, which spans exon 1 through exon 3. This genomic region was investigated by Kusuietal. Methylation effects on gene expression are indicated by the narrow box. The MT2 region in particular (shown in black) was shown to be functionally significant for OXTR gene regulation. Panel B shows the regions within the CpG island that were investigated with regard to differential methylation. Filled circles indicate individual CpG sites with significant differences in methylation levels. Numbering is relative to the translation start site (+1).

Source: Adapted from Kumsta et al. (2013).

to play and sport. The imprinted genes are not permanent; the emphasis on silencing gene expression by imprinting or emphasizing is, of course, an active area of research in brain functioning (Crews 2011).

Figure 4.6 depicts epigenetic events: oxytocin receptor and oxytocin gene expression is tied to social behavior and most probably underlies certain social sports. Such events in biological structure underlie sports capability and the shaping of the brain through training and practice. For instance, genes that produce oxytocin expression are particularly significant. This important peptide hormone plays diverse regulatory roles, from lactation to parturition to social attachment, depending on where it is expressed in end organ systems (Carter, Lederhendler, and Kirkpatrick 1999).

As research continues, we should be able to see what types of genomic receptor variation take place in those demonstrating superior skill in, for example, basketball shots, hockey finesse, or skiing expertise. That is still the marvel of it all: that someone can actually do the sorts of things that we witness in Olympic competitions, professional games, and amateur athletic events.

PRACTICE

Genetic talent and epigenetic change are put into action by practice. There are few great athletes, and there are many good ones, and their common feature is practice, practice, practice, in the development of skill. Practice allows us not to have to think so hard about what foot goes where in executing a double axel rotation in figure skating or about how we should be standing when we are trying to sink a golf ball. It's not that we give up on conscious thinking—the skater is still declaratively aware that she is going to execute a double axel—but constant practice allows her to rely on muscle memory and unconscious knowledge to execute it, which frees up her conscious memory to deal with variables such as the condition of the ice, the bouquet of flowers lying on the rink, and the reaction of the audience. Without practice, it is possible to become frozen in thought or immobilized by overthinking, and this is ineffective in terms of action response.

Sport reminds us why practice, practice, and more practice in learning the form is essential to the development of sport excellence: shoot for the hoop over and over again, so much so that it becomes embedded in one's head, like it has been for Stephen Curry, the 2015 NBA Most Valuable Player. Shooting from all sides and angles, running and darting about—in that way the continuity of the ball, the detection and computation of the angle and terrain, become one in the moment of performance.

Without practice, some people can become good enough at what they do, but very few become great. Indeed, we know that the development of skill "sculpts" the neuronal response in the primary motor cortex (Matsuzaka, Picard, and Strick 2007); in other words, it actually changes the brain. What is apparent in the evolution of the cortex is first the expansion of the cortex in general, including the motor cortex and the very pronounced visual cortex in particular. What pervades the motor systems are diverse anticipatory and cognitive systems that figure throughout action in general and sport in particular, for which information molecules such as dopamine are fundamental.

Most children *play* games, but those who focus on one game in particular may end up being sportspeople. Many people discuss the differences in terms of practice. So an amateur *plays*; a pro *practices*. But consider how many athletes excelled at multiple sports in high school and college (Allen Iverson, Russell Wilson) but were forced to focus on one because playing a professional sport requires year-round attention. Very few have successfully played two sports at a professional level (examples are Clara Hughes, Bo Jackson, and Deion Sanders).

The interaction of genes and social context are clearly important to sports development. The plasticity built into pedagogy and cephalic function turns raw ability into practiced skill. And skill on the sports field may translate into other fields. Studies show that diverse kinds of athletes are better than the average person at cognitive experiments tied to anticipation of movement and requiring speed and accuracy (Erickson et al. 2011).

Practice, genetic capability, and social context feed the continuity of biological and cultural evolution, shaping cephalic function and facilitating further excellence of bodily sensibility. Genomewide and specific changes in neural function are commonplace across pedagogy and development; indeed, perception and action are joined at the hip in the excellence involved in adaptation, and perception and action underlie all of sport and human life.

CONCLUSION

We live in the age of biology, of broad-based discoveries such as the decoding of the genome. This has included some odd surprises. For instance, the number of genes in our species is much smaller than what we expected, especially from an evolutionary point of view.

We now know the genomic makeup of our species and that of many others. But we know little about how the assemblage of genes works as a whole. A lot of structure has been revealed, but we have yet to find out much about how it all is linked. Our understanding of sports is hampered by our current state of knowledge.

Nevertheless, genetic and epigenetic studies have helped us understand features such as endurance, learning capability, and temperament and motivation variables. We know that perception, strength, pain capability, and focus matter in sport. It is these small components that cohere into something we might call excellence in sport.

Anticipatory ability is pervasive in bodily regulation and in sports, for both geniuses and the less-than-genius. Most systems are action oriented and motivated, sampling and corrective systems dominate the regulation, and the systems are fast. And minimizing damage is a constant. Athletes run through lower-level physiological functions within appraisal systems concerning space, time, and probability. The computational systems used in sports are phylogenetic adaptations in our species. The assessment of objects in space, tied to time and mined by probability, traverse diverse regions of the brain. And this is just not a cortical affair.

Indeed, subcortical regions such as the amygdala and hippocampus are critical for computing the location of objects in space, anchored to time and knotted to prediction; so are many neocortical regions. Sports excellence is about adapting to change (Sterling and Laughlin 2015)—and sports excellence reveals the expansion of skills embedded within biological capability and enhanced through pedagogy, invention, and innovation.

In the next chapter, we will turn to long-term viability, coping with recovery, and resilience in sport.

5

REGULATION, RECOVERY, AND RESILIENCE

To participate successfully in sports, the human body—and I include in that body the human brain and mind—needs to be able to manage or regulate its internal milieu, recover from injury, and show resilience in the face of stress and strain. We all know of figures in sport for whom recovery from injury is an essential part of their sporting history; at times the recovery process draws on the same themes of recovery as appear in ontogeny or development. Age and genetics, in addition to discipline and temperament, figure in the capacity for recovery in sports.

Sport is one vehicle among others to facilitate adaptation to adversity and to promote a sense of resiliency (Bell and Suggs 1998). Learning about pain and recovery goes with the turf of being human: battling fear, coping with misfortune. Sports, therefore, mimics broad-based human adaptation.

In this chapter, I begin with emotion and sport and then describe some of the bodily regulation that figures importantly in sport and the adversity or stress that underlies sports events. Coping and adaptation are inherent across the broad array of competitive sports. Biological capabilities made this possible, and sport enhances both mental and physical capabilities.

EMOTION IN SPORT

Our emotions are crucial survival mechanisms (Darwin 1872, LeDoux 2012). Their regulation and integration into our plan of action is key, as they highlight ends that must be satisfied (Dewey 1895). They can be adaptive or not, but they are essential to sport.

As the television show *Wide World of Sports* used to put it, athletics is full of "the thrill of victory and the agony of defeat." Emotions run high in sport, and indeed they are essential for the good, the bad, the elated, and the defeated. The machinery of the brain is rich in emotional appraisals. The emotions are great sources of information, certainly for conspecific contact and for relaying information to others, in competition or not. Emotions can be disruptive in general and in sport in particular, which is why we often talk of the need to control them. The issue is their regulation and the activation of them in the context of adaptation.

Darwin (1871) emphasized a fundamental prosocial human feature, one essential for moral judgment. He asserted: "Any animal whatever, endowed with a well-marked social instinct, the parental and filial affections being here included, would inevitably acquire a moral sense of conscience, as soon as its intellectual power had become as well or nearly as well developed as in man" (95). But in *The Descent of Man*, Darwin (1871) also noted that the "the fewness and the comparative simplicity of the instincts in the higher animals are remarkable in contrast with those of the lower animals" (65). Figure 5.2 presents a striking depiction of Darwin

FIGURE 5.1

Human cephalic metaphor.

FIGURE 5.2

Facial expressions from Darwin's book on emotions.

Source: Darwin (1872).

seasoned by age. Also depicted in these figures are the facial expressions described in Darwin's 1872 book on the biology of communicative expressions; that book makes reference to diverse forms of facial responses and our response to them in the organization of action.

As the emotions traverse our experiences in sports, so do moral appraisals; these are tied to empathy and other promoral sentiments. Of course, nineteenth-century theorists such as Charles Darwin (1872), William James (1890), and John Dewey (1895) anchored the emotions in adaptation. They did not set up a dualism between reason and emotion. Emotion pervades sport as it does the rest of life.

Emotions must be inhibited in some sports (competitive rifle shooting, for instance, where control of breathing and heartbeat are crucial) and overly activated in others (such as ice dancing, which requires consider-able emotional expression for top scores). But emotions are key to our survival as a species. They evolved to aid adaptation, to indicate when to approach and when to avoid, to help us seek comfort and to threaten when necessary. They are part of the information-processing systems we use to discern events. Within the neural systems are diverse information molecules that underlie the emotions across the neural axis, including visceral peripheral systems (primarily, the gastrointestinal tract).

Appraisal systems pervade emotional detection and expression, and they are tied to survival systems. While in the Western world detached reason has often been highlighted as the paragon of rationality and emo-tion as something that denigrates reason, emotion and cognition are intertwined throughout the brain as part of survival systems (Lazarus 1991; Parrott and Schulkin 1993; LeDoux 2000, 2012, 2016). Emotions are no more or less adaptive than other cephalic features; the issue involves regulation and flexibility suited to context and prediction, all of which are inherent across sport.

Sport is rife with coping and surviving, as is most human activity. Ath-letes can cope with problems by utilizing behavioral strategies (increas-ing efforts, following routines, taking advice, technique-oriented coping, changing behaviors, communicating) or cognitive strategies (increasing attention, focusing on one's role, using mental imagery and instructional self-talk, planning). Similarly, they can cope with negative emotions by relaxation, reappraisal, external verbal persuasion, mental imagery, self-talk, and blaming. An athlete copes with avoidance by either physically withdrawing from the situation that causes the need to avoid or by block-ing the thoughts related to avoidance (Campo et al. 2012).

Sport is about winning; contact sports are easy to read. There are clear emotional antecedents in contact sports. Feelings of health are predicated on actual physical well-being, injury, and pain. The emotional influence of others comes from negative aspects of relationships, conflicts, criti-cisms from others, referees' wrong decisions, opponents' cheating, oppo-nents' performance, and teammates' behavioral events. Emotions while playing also are derived from noncompetitive antecedents: diet, weather, sleep, organization, career, and work and work/nonwork interface (Campo et al. 2012).

Thus, emotion fuels sport on all levels, perhaps especially so when playing at home, as seen in the common concept of the home-field advantage. Though there should realistically be no difference playing on identical courts/fields in different locations, the emotion, exhilaration, and adrenaline of playing for a crowd on one's own side can make a large difference.

FOOD AND SPORT

Food is a major source of the nourishment required to regulate, rebuild, and repair our physical systems. And, of course, food has social and cultural meanings. Knowledge acquisition is also embedded in our omnivorous appetitive and consummatory experiences (Craig 1918, Dewey 1925). The range of objects we humans eat is vast. Brains, for instance, are consumed in all human cultures—including, in some places, human brains.

Not all of our omnivorous tendencies have such unfortunate outcomes as cannibalism. However, we do come prepared to symbolize core organs in mythological expression—for example, the dead brain giving life to the living. In our cultural milieu, consider the range of products that claim

FIGURE 5.3

The brain as food: calf brains sold by the pound.

to boost brain function: so-called brain food. (Then again, what does *not* affect the brain?)

Another side of omnivory is the wide range of objects we consume. In prehistoric times we banded together to hunt, foraging for food resources and taking advantage of the great array of plants, fruits, vegetables, water, and minerals available to us.

A number of beverages are currently available that can "boost brains." For example, NeuroSonic, Brain Toniq, and the "Focus" flavor of Vitamin Water are stimulants that provide caffeine and L-theanine, eluthero root and rhodiola root, and vitamin A, respectively, to improve mental or physical functioning. Whereas NeuroSleep, Slow Cow, and Marley's Mellow Mood, on the other hand, are quiescents that use melatonin, L-theanine, valerian root, and chamomile to calm the mind and body.

Besides brain food, consider the modern ingesta available to enhance performance in diverse sports. General Nutrition Centers (GNC), which sells health and nutritional substances, has a special page for sports nutrition, divided into six areas (protein, preworkout, postworkout, hardcore, thermogenics, and creatine). The preworkout section alone offers one hundred and twenty-two items, including such products as Cellucor C4 Extreme, BPI-Pump HD, and BullNOX Androrush, all in a variety of flavors. Whether these items are more effective or safer than your auntie's brains is an unanswered question.

Over billions of years, the search for food resources, fluid, sexual contact, and social connection has informed our evolutionary background. As hominins developed cultures, they also created symbolic rites that fostered social contact. It would not have taken long from there to develop folk knowledge. For example, damage to the head is debilitating or even deadly. Play and then sport, as we evolved, provided safe(r) outlets for expanding capability and facilitating social contact.

And it must have been obvious as well how important the heart is to life. Eating or pulling out the heart—the source of life—from live humans was a feature of rites of sacrifice and power in some cultures. This enacted the giving of life to obtain more; ability along with practice was translated into hunting capability, fertility, and eventually sport, or the mythology of competence in sport.

REGULATION OF THE INTERNAL MILIEU

Diverse anticipatory systems activate when we are about to ingest a food resource. This includes the rich array of information molecules expressed in the brain and peripheral nervous system in the absorption and use of food resources (Richter 1943, Swanson 2000, Moran 2000, Power and Schulkin 2009). For instance, peptide hormones, which are expressed in both the peripheral and central nervous systems, work in the context of the larger physiological and behavioral control of food ingestion (Moran 2000).

Pavlov called this the "cephalic phase": anticipatory systems in the utilization of resources required by the body, the secretion of insulin before the utilization of the resource (Richter 1943, Todes 2014), and the secretion of insulin based on the time of day in which a food source might be available (Powley 1977). Thus, simply the *anticipation* of food generates some of the many information molecules vital for the acquisition and utilization of food resources. These events are cephalic, although the context of food passing through the oral cavity also facilitates the anticipation of utilizing and distributing vital food resources.

The central nervous system interprets this information within the constraints of experience (knowledge); intrinsic, evolved tendencies (phylogeny); and current conditions (for example, social setting, nutritional status). Is there a threat? Is there an opportunity? The central nervous system sends messages to the appropriate peripheral organs to begin the physiological cascades that prepare the organism to respond to the anticipated challenge. These events underlie long-distance sports and the secretion of information molecules to maintain viability. Thus, a modern-day human, rather than going out and killing breakfast, has the energy for a two- or three-mile morning run.

Cephalic-phase and anticipatory digestive and metabolic responses to cues to imminent feeding allow organisms to get "ahead of the curve." They improve the efficiency with which animals digest food and then absorb, metabolize, and store the liberated nutrients. They also prime the organism to meet the resulting challenges presented by the influx of nutrients, such as changes in blood pH and electrolytes. Athletes' bodies also adapt to training and learn to get better and better at regulation.

Multiple mechanisms in the brain provide the potential for cephalic anticipatory adaptation—that is, the brain constantly attempts to anticipate future events. This is paramount for the functioning of complex social systems, and adaptation is achieved, in part, by the cephalic regulation of behavioral and physiological systems in the expression of longer-term adaptation (Sterling and Eyer 1988, Schulkin 2003).

Traditional conceptions of regulation have typically (though not always) emphasized homeostatic short-term regulation, with a further focus on set-point stability and short-term contemporaneous adaptation (Cannon 1916, Dallman and Bhatnagar 2000). The expansion of cortical function, in social groups of diverse complexity, entails longer-term regulation.

The emphasis is on social competence (which would include cooperative behavior in tool use) but, since we are social animals, also on an evolved set of anticipatory mechanisms that facilitate the regulation of the internal milieu amid an expanding social milieu. And these behavioral adaptations impact the brain directly; evolution selected for behavioral adaptations to feed back directly into cephalic function.

Cephalic machinations integrate our internal physiology via behavioral regulation (Richter 1943), mediated by the rich innervation of peripheral sites (Powley 1977, 2000; Power and Schulkin 2009) and the activation of peripheral sites by vagal efferent projections into sites along the digestive tract (Powley 2000).

Importantly, cortical sites (for example, the frontal cortex) project to these brainstem sites, expanding cortical function into larger forms of visceral control and revealing diverse routes of anticipatory regulation of the internal milieu essential for sport. Regulation of the internal milieu is linked to the social environment, a connection that succeeds or fails in the balance between cooperation and competition for resources.

Glucose is a prime example of a nutrient whose blood concentration is actively regulated by both reactive and anticipatory physiology. If blood glucose falls below a critical concentration, it can lead to rapid brain damage and then death. However, high concentrations of blood glucose are potentially toxic and are associated with macular degeneration, brain cell death, and higher mortality after stroke. A number of mechanisms have evolved to resist changes in blood glucose concentration and keep it within safe levels. Glucose is constantly being shuttled among different pools, or, perhaps more precisely, the energy contained in glucose is shuttled among these pools.

Genetic differences underlie body regulation during exercise, including oxygen uptake (Bouchard et al. 2000), heart rate (cAMP-responsive element-binding protein 1), glucose regulation (peroxisome proliferator-activated receptor), and fat-mass regulation (for example, FTO genotype) (Bouchard et al. 2000). The regulation of glucose and insulin must be set in the context of the genetics that underlie muscle adaptation (Phillips et al. 2013).

Insulin is the primary peptide regulating glucose metabolism. Insulin increases glucose storage (in the form of glycogen) in liver and muscle, decreases lipolysis and gluconeogenesis, and increases fatty-acid synthesis by adipose tissue. The net result is lowered blood glucose concentration through an increase of the conversion of glucose to other energy-storage molecules (glycogen and fat) and a decrease in the production of glucose by the liver.

Anticipatory physiological regulation is an adaptive strategy that enables animals to respond faster to physiologic and metabolic challenges. The cephalic-phase responses are anticipatory actions that prepare us to digest, absorb, and metabolize nutrients essential for sports. Cephalic-phase responses are a fundamental concept in regulatory physiology and a prime example of anticipatory physiological responses. They need to be integrated across the diverse regulatory information molecules that are being discovered and viewed in an adaptive, evolutionary perspective.

Steroids, such as the adrenal steroid cortisol, can have opposite effects from their originating purpose when elevated in the brain for long periods. In the short term, cortisol is essential in maintaining behavioral and physiological functions for energy regulation. Glucocorticoids are essential for glucose metabolism for activity in general and sport in particular. Long-term elevation can be harmful, however, as noted by the breakdown of diverse tissues in such cases (Sapolsky 1992; Dinkel, Dhabhar, and Sapolsky 2004).

Glucocorticoids are primarily related to the organization of glucose metabolism, something in which all cells are implicated. They participate with metabolic hormones (for example, insulin) in the regulation of metabolism and in the generation of searching and feeding behavior

(Dallman 2003), in part by activating diverse neurotransmitters and neuropeptides (Woods, Hutton, and Makous 1970). In fact, diverse forms of appetitive behaviors are associated with elevated levels of cortisol (Schulkin 1999, 2003; Dallman 2003). They are linked to diverse forms of foraging behaviors, sports, and anticipation of rewards (Wingfield et al. 1999).

Moreover, glucocorticoids vary their effects seasonally, during droughts, and in general in prediction of the availability and allocation of food resources and other metabolic requirements (Wingfield et al. 1999). Glucocorticoids are not a one-dimensional entity. They are multidimensional in action and effect and thus fundamental to biological and social adaptation (see table 5.1).

Thus, glucocorticoids promote foraging behavior in many different animals under various conditions, and they facilitate life processes essential for reproduction, energy metabolism, and attention to external events (Sapolsky 1992, Schulkin 2003). The important point in understanding regulatory physiology's program of promoting viability over time is the adaptive nature of various hormones' actions depending on circumstances both internal and relative to niche and circumstance (Wingfield et al. 1999, Adkins-Regan 2005).

Cortisol, long considered "the hormone of stress," is mischaracterized when understood only in that way. It is the hormone of energy metabolism and adaptation through which different end organ systems are mobilized or restrained for action (Schulkin 2003, Sapolsky 1992, McEwen 2008). The secretion of cortisol is an adaptive response serving physiology and behavior. The continued secretion of cortisol without relief has consequences,

TABLE 5.1 Short-term adaptations and long-term detriments of glucocorticoids

Short-term adaptation	Long-term disruption
Regulation of immune system	Disruption of immune response
Increase glycogenesis	Protein loss
Increase foraging behavior	Growth and reproductive disruption
Increase activation of the brain	Bone-mineral loss
	Brain deterioration

Source: Adapted from Wingfield and Romero (2001).

though: bone demineralization, compromised immune system, and shifts in metabolism (Sapolsky 1992; Dinkel, Dhabhar, and Sapolsky 2004). One important adaptation is regulation of the internal milieu—that is, the secretion of cortisol—through manipulations in and changes to the social context.

But this breakdown does not have to be permanent: it is recoverable, though age dependent. It is harder to recover from injury when one is forty years old than when one is ten or twenty, perhaps because of neural deterioration as a function of high levels of cortisol (Sapolsky 1992; Dinkel, Dhabhar, and Sapolsky 2004). This plasticity or change is tied to sustaining tissue and then regulating the tissue to viability and longer-term consideration (Power and Schulkin 2009). This regulation is essential in sports (especially running) and in activity and life in general.

INJURY AND RECOVERY

Many of our adaptive responses are oriented toward protecting ourselves from danger and injury. This is outstandingly clear in sport. The body is designed to detect danger and institute protective responses when something impinges and creates vulnerabilities.

Brain damage in particular raises fears in many sports. Football players, given their many impacts on the field, are particularly prone to concussion. But baseball and tennis players get hit in the head with balls; mountain climbers suffer from falls and hypothermia; cyclists, bobsledders, and lugers can fly off their bikes and sleds at potentially fatal speeds. Interestingly, though, certain life events and experiences that occurred before the brain damage, such as the practice of sport, can moderate some of the debilitating effects. In a variety of experiments, motivation of hunger and weight loss has been shown to adjust some of the metabolic and feeding disorders that occur from lateral hypothalamic damage. The same holds for alimentary experiences (for example, salt hunger and thirst).

The debilitating sensorimotor effects that can result from brain damage are decreased through such alimentary experiences, as are visceral lesions. Recovery of function is knotted to diverse experiences; for hippocampal damage, various memory experiences decrease some of the memory decrements. Damage to other regions of the brain, such as the septum, augment rage and aggression.

It has long been known that environmental enrichment of experience increases neuronal plasticity. For instance, young rats given larger and more objects to play with have greater synaptic hybridization across several brain regions. The expansion and enrichment and plasticity of the brain can occur from both diverse preoperative events and postoperative experiences (Adkins et al. 2006), and diverse genomic changes are fairly commonplace as a result (Rampon et al. 2000).

One feature is motor enrichment. Since Sir Charles Sherrington (1906), it has been long noted that reduced food access results in brain excitability in the motor cortex. Running activity, for instance, enhances the production of diverse information molecules (more on this below) including neuronal growth factors and dopamine.

Recovery from brain damage is tied to the overactivation of the remaining neuronal subset. Dopamine is a good example; lateral hypothalamic damage, which severely undercuts motor control, results in the overactivation of the remaining dopamine neurons in this region, compensating for the loss and aiding the recovery from damage to this region. Synaptic branching is another mechanism of recovery; it is implicated in learning and in the recovery of function and variation as well as functionally relevant forms of adaptation (Adkins et al. 2006).

Certain forms of trauma to the head may not reveal themselves immediately but eventually become obvious; we are familiar with such events in the lives of boxers and football players. Pre-expression of brain damage before it is obvious may manifest as a form of PTSD. An infusion into the ventricle that does not result in any observable seizure can nevertheless facilitate simulated cocaine-induced seizures. Glucocorticoids may potentiate such effects.

In an evidence-based analysis, diverse risk factors are operative. Of these, prior concussion is a major factor. Men appear to be at increased risk when they participate in alpine sports, American football, and lacrosse; women who play soccer are at risk, and so are cheerleaders. Finally, one gene may be linked to concussions: apoilipoprotein E (APOE), a lipid-related molecule.

TABLE 5.2 Histopathological classification of modern chronic traumatic encephalopathy

BU CSTE CRITERIA: FOUR STAGES OF DISEASE[a]	FOUR PHENOTYPES[b]
Stage 1: Normal brain weight. Focal epicenters of perivascular p-τ and NFTs and astrocytic tangles involving the sulcal depths and typically affecting the superior and dorsolateral frontal cortices.	Phenotype I: Sparse to frequent NFTs and neuritic threads in the cerebral cortex and brainstem, without involvement of the subcortical nuclei (basal ganglia) and cerebellum. No diffuse amyloid plaques in the cerebral cortex.
Stage 2: Normal brain weight. Multiple epicenters at the depths of the sulci, with localized spread from epicenters to the superficial layers of the adjacent cortex. No NFTs or p-τ involvement in the medial temporal lobe.	Phenotype II: Sparse to frequent NFTs and neuritic threads in the cerebral cortex and brainstem, with or without such pathology in the subcortical nuclei (basal ganglia) and cerebellum. Diffuse amyloid plaques in the cerebral cortex.
Stage 3: Mild reduction in brain weight. Mild cerebral atrophy, with dilation of the lateral and third ventricles. Septal abnormalities. Moderate depigmentation of the locus coeruleus and mild depigmentation of the substantia nigra. Atrophy of the mamillary bodies and thalamus. Widespread p-τ pathology in the frontal, insula, temporal, and parietal lobes. NF pathology in the amygdala, hippocampus, and entorhinal cortex.	Phenotype III: Brainstem predominant: moderate to frequent NFTs and neuritic threads in the brainstem nuclei; absent or sparse NFTs and neuritic threads in the cerebral cortex, subcortical nuclei (basal ganglia) and cerebellum. No amyloid plaques in the cerebral cortex.
Stage 4: Marked reduction in brain weight, with atrophy of the cerebral cortex. Marked atrophy of the medial temporal lobe, thalamus, hypothalamus, and mamillary bodies. Severe p-τ pathology affecting most regions of the cerebral cortex and the medial temporal lobe, sparing the calcarine cortex. Severe p-τ pathology in the diencephalon, basal ganglia, brainstem, and spinal cord. Marked axonal loss of subcortical WM tracts.	Phenotype IV: Absent or sparse NFTs and neuritic threads in the cerebral cortex, brainstem, and subcortical nuclei (basal ganglia). No cerebellar involvement. No diffuse amyloid plaques in the cerebral cortex.

Notes: LC = locus coeruleus; NF = neurofibrillary; NFTs = neurofibrillary tangles; p-τ = phosphorylated τ; SNr = substantia nigra; WM = white matter.

Source: Adapted from Gardner et al. (2013); Omalu et al. (2005).

But what is a concussion? Table 5.2 lists some of its features. Concussions are generally rated as mild, moderate, or severe. There are no conclusive monolithic features, and damage can range from almost imperceptible to severe or traumatic (Blennow, Hardy, and Zetterberg 2012). Some effects of concussions include a headache and a foggy feeling, labile emotions and irritability, unconsciousness, amnesia, slowed reaction time, and sleep disturbance (McCrory 2009).

Recovery from concussion is equally variable. Figure 5.4 shows one set of recovery features derived from college football players with mild concussion. Glucose metabolism is accelerated under conditions of adaptation in action, and it is similarly exaggerated during recovery from brain damage or concussion. But this has its price: further deterioration can be triggered, and recovered athletes can be more susceptible to neural vulnerability and damage.

Aside from brain damage, athletes are prone to other physical injuries in games as well as during practice, although fewer injuries occur at practice (see figure 5.5 and Table 5.3). Sports injuries may result from acute events—a blow from a baseball breaking a bone in the forearm or a long-distance jumper snapping an Achilles tendon—or from long-term stress and strain. Figure skaters, for instance, such as the great Russian skater Evgeni Plushenko, are prone to back injuries after years of repeated jumping, rotating, and landing on hard surfaces. Plushenko has had multiple back and knee surgeries. While he was able to return to competition and won a gold medal skating in the Olympic team competition in 2014, a new back injury forced him to pull out of the men's singles competition.

But while athletes are more likely to suffer injury than desk jockeys, they are also more likely to recover quickly, even from life-threatening injuries. The tennis great Arthur Ashe, who won at Wimbledon in 1975, suffered a massive heart attack in 1979 and was forced to undergo a quadruple bypass (Ashe and Rampersad 1994). His disease was likely the result of hereditary heart disease, but his superb physical condition helped him survive the ordeal and recover remarkably quickly. (Unfortunately, he also acquired HIV from a blood transfusion during a corrective bypass surgery and died in 1993 from AIDS-related pneumonia.)

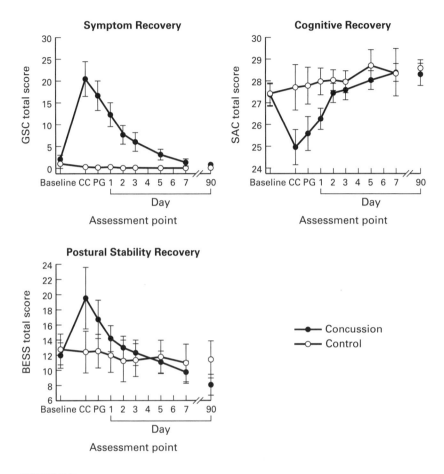

FIGURE 5.4

Higher scores on the Graded Symptom Checklist (GSC) indicate more severe symptoms, lower scores on the Standardized Assessment of Concussion (SAC) indicate poorer cognitive performance, and higher scores on the Balance Error Scoring System (BESS) indicate poorer postural stability. Error bars indicate 95% confidence intervals. CC indicates time of concussion; PG, postgame/postpractice. On the BESS, multiple imputation was used to estimate means and 95% confidence intervals for control participants for the time of concussion and postgame/practice assessments.

Source: McCrea et al. (2003).

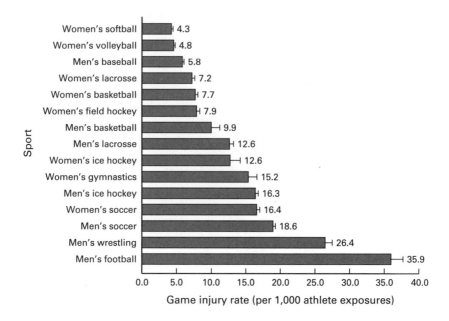

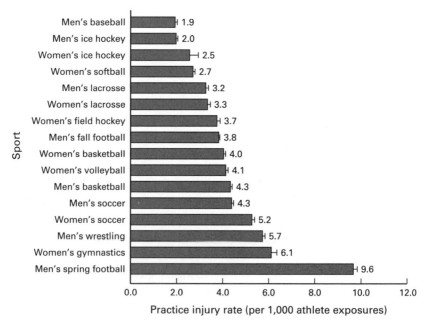

FIGURE 5.5

Injury rates among athletes.

Overall game (top) and practice (bottom) injury rates for fifteen sports, National Collegiate Athletic Association, 1988–1989 to 2003–2004. Although data for fifteen total sports are presented, fall and spring football are reported separately for practices. Because no "official games" are played during spring football, only fall football is listed for games.

Source: Hootman, Dick, and Agel (2007).

TABLE 5.3 Injuries in collegiate sports

INJURIES	FREQUENCY	PERCENTAGE OF ALL INJURIES	INJURY RATE PER 1,000 ATHLETE EXPOSURES
Ankle ligament sprains			
Men's baseball	663	7.9	0.23
Men's basketball	3,205	26.6	1.30
Women's basketball	2,446	24.0	1.15
Women's field hockey	327	10.0	0.46
Men's football	9,929	13.6	0.83
Women's gymnastics	423	15.4	1.05
Men's ice hockey	296	4.5	0.23
Women's ice hockey*	12	2.8	0.14
Men's lacrosse	698	14.4	0.66
Women's lacrosse	602	17.7	0.70
Men's soccer	2,231	17.2	1.24
Women's soccer	1,876	16.7	1.30
Women's softball	526	9.9	0.32
Women's volleyball	1,649	23.8	1.01
Men's wrestling	715	7.4	0.56
Men's spring football	1,519	13.9	1.34
Total ankle ligament sprains	27,117	14.9	0.83
Anterior cruciate ligament injuries			
Men's baseball	56	0.7	0.02
Men's basketball	167	1.4	0.07
Women's basketball	498	4.9	0.23
Women's field hockey	53	1.6	0.07
Men's football	2,159	3.0	0.18
Women's gymnastics	134	4.9	0.33
Men's ice hockey	78	1.2	0.06
Women's ice hockey*	3	0.7	0.03
Men's lacrosse	131	2.7	0.12
Women's lacrosse	145	4.3	0.17
Men's soccer	168	1.3	0.09
Women's soccer	411	3.7	0.28
Women's softball	129	2.4	0.08
Women's volleyball	142	2.0	0.09
Men's wrestling	147	1.5	0.11
Men's spring football	379	3.5	0.33
Total anterior cruciate ligament injuries	4,800	2.6	0.15

(continued)

TABLE 5.3 (*Continued*)

INJURIES	FREQUENCY	PERCENTAGE OF ALL INJURIES	INJURY RATE PER 1,000 ATHLETE EXPOSURES
Concussions			
Men's baseball	210	2.5	0.07
Men's basketball	387	3.2	0.16
Women's basketball	475	4.7	0.22
Women's field hockey	129	3.9	0.18
Men's football	4,404	6.0	0.37
Women's gymnastics	64	2.3	0.16
Men's ice hockey	527	7.9	0.41
Women's ice hockey*	79	18.3	0.91
Men's lacrosse	271	5.6	0.26
Women's lacrosse	213	6.3	0.25
Men's soccer	500	3.9	0.28
Women's soccer	593	5.3	0.41
Women's softball	228	4.3	0.14
Women's volleyball	141	2.0	0.09
Men's wrestling	317	3.3	0.25
Men's spring football	612	5.6	0.54
Total concussions	9,150	5.0	0.28

* Data collection for Women's Ice Hockey began in 2000–2001.

Source: Adapted from Hootman, Dick, and Agel (2007).

TABLE 5.4 Common sports injuries and their treatments

INJURY TYPE	MAIN SPORTS	TREATMENT(S)
Ankle		
Lateral ligament injuries, medial ligament injuries	Soccer, basketball	Reconstruction of ligament with a peroneal tendon or fibular periosteal flap
Pott's fracture	Basketball, volleyball	Immobilization in a cast
Knee		
Anterior cruciate ligament (ACL) injury, posterior cruciate ligament (PCL) injury	Soccer, skiing, American football, basketball	Graft from patella, hamstring, or quadriceps, or from a cadaver
Meniscal tear	Soccer, skiing, American football, basketball	Arthroscopic surgery to repair or trim the torn part of the meniscus
Runner's knee	Cross-country, track and field	Arthroscopic surgery to remove fragments, or surgery to realign kneecap

TABLE 5.4

Hand/Wrist		
Sprains, fractures	Various	Immobilization
Dislocation of the proximal interphalangeal joint (PIP) joint	Contact sports	Immobilization
DeQuervain's syndrome	Tennis, racquet sports	• RICE* and immobilization • In some cases, cortisone injections • In rare cases, surgery to release the tendon sheath
Extensor carpi ulnaris tendinitis	Basketball, racquet sports, golf	As above
Baseball finger	Baseball, softball	• RICE and immobilization • In some cases, surgery using pins, wires to secure bone fragments and realign the finger
Elbow		
Ulnar collateral ligament injury	Baseball	"Tommy John surgery": a tendon from the arm, patella, or from a cadaver replaces the torn or injured ligament
Shoulder		
Dislocation	American football, contact sports	• Generally, external manipulation of the shoulder to realign the joint using techniques such as the scapular manipulation or Hennepin maneuver • At times, a physician may need to perform an open reduction in which the shoulder is realigned during an open surgical procedure
Impingement	American football, contact sports, weightlifting	Anti-inflammatory medications, reduced movement, strengthening exercises
Rotator cuff injuries	Tennis, baseball, freestyle swimming	Surgery to reattach the tendon to the head of the humerus

*The first-line (and preferred) treatment for many of these conditions is RICE (Rest Ice Compress Elevate). Surgery is usually necessary only in the most extreme cases.

Source: Adapted from Brukner and Khan (2006); Doral et al. (2012); Walker (2007).

RESILIENCE

Athletes such as Ashe, Plushenko, and Kerri Strug, despite their serious injuries, demonstrate resilience. The term "resilience" is usually applied to individuals who experience lasting stress or trauma but continue to be productive and healthy. Resilience describes a trajectory of responses to duress that do not include periods of significant ill health or decline and that represent a good-enough outcome (McEwen 1998). Resilience is the ability to continue to function within normative boundaries while undergoing extreme changes and inconsistencies. Treatments for injuries complement the athlete's mental and physical resilience through the injury's duration. Table 5.4 lists common injuries in sport and the available treatments for these injuries. In addition, a person's ability to anticipate future, more stable circumstances may allow him or her to maintain relative stability in the face of dramatic changes. Apart from appearing in many psychological contexts, resilience shows many neurobiological expressions.

Importantly, levels of cortisol following a traumatic event may play a role in a person's ability to express resilience (McEwen 1998). A number of mediators to the effects of cortisol have been proposed, among them brain-derived neurotrophic factor (BDNF) in the hippocampus (Karatsoreos and McEwen 2011). While high levels of BDNF have been linked to depressive symptoms, findings indicate that a reduction in BDNF in response to stress, particularly stress experienced at a young age, may reduce the expression of resilience. Such factors also affect sport capabilities.

Childhood trauma has been linked to neurobiological changes that encourage bodily breakdown and therefore diminish resilience (Grassi-Oliveira and Stein 2008). In animal studies, rats with a history of maltreatment showed signs of hyperresponsiveness. Human studies also demonstrate that individuals exposed to trauma at a young age show increased stress responses in the Hypothalamic-Pituitary-Adrenal (HPA) axis when compared with those with no history of trauma (Davidson and McEwen 2012). Because of this, a person is likely to experience increased levels of cortisol for an extended period of time, decreasing chances of resilience. Multiple studies also point to structural changes in the brain as a result of childhood trauma, most typically an enlarged amygdala and decreased volume in areas of the prefrontal cortex (Davidson and McEwen 2012). These early neurobiological changes may result in social changes that affect a person's response to his or her environment in ways that also decrease resilience.

Adult chronic stress has also been linked to multiple physiological changes that disrupt stress regulation. Elliot and colleagues (2010) found that stressed adult mice that displayed symptoms of defeat (for example, social withdrawal) showed higher levels of demethylation and CRH. This evidence supports the hypothesis that demethylation and higher levels of CRH may lead to stress-induced psychopathology.

Some neurochemical factors may encourage resilience and act as protective factors. Morgan and colleagues (2004) found that higher levels of dehydroepiandrosterone (DHEA) during military training are associated with better performance under stressful circumstances. According to these studies, higher levels of DHEA appear to counteract some of the deleterious effects of elevated cortisol and increase resilience in stress-inducing situations. Neuropeptide Y (NPY) seems to balance the effects of CRH in ways that may encourage more resilient responses to stressors. These and other factors are being explored for possible pharmacological interventions for people who encounter adversity. It is possible that by increasing neural plasticity through medical or psychotherapeutic means, for instance with DHEA or NPY, individuals in a state of overload could experience recovery and those at high risk may develop more adaptive pathways (Karatsoreos and McEwen 2011).

Additionally, these internal factors inform our social behaviors and modify the way that an individual relates to the world around him. The interplay between all of these factors determines a person's ability to anticipate and react to adversity in an adaptive way.

That social contact and experiences directly affect our physiological state and cognitive processes is an impressive testament to the importance of social context to human survival (Schulkin 2011). Humans have evolved a "cognitive penchant" for considering the long-term implications of their behaviors toward others, such as the costs and benefits of social cooperation and social knowledge (Foley 1996). Humans are also unique in the degree to which they engage in cooperative behaviors in order to achieve common ends. The ability to share common goals and to anticipate, account for, and consider others may be among the greatest human cognitive adaptations.

These social contacts build in safeguards in anticipation of future duress and help determine whether an individual will experience resilience.

An abundance of evidence demonstrates the importance of interpersonal interactions and social support in reducing the impact of stress and substantially increasing neurobiological flexibility. Coan, Schaefer, and Davidson (2006) found that among women exposed to the threat of shock, those who held a partner's hand exhibited lower stress responses than those holding hands with a stranger or with no one. Strong social bonds and social support can potentially prevent or reduce some maladaptive responses to life stressors. Davidson and McEwen (2012) identify several treatments that increase social skills, resulting in improved positive affect and changed brain functions associated with plasticity.

Increased levels of cortisol, however, can devolve from social functioning, reducing social meaning and social contact, which normally act as ameliorative factors. Group affiliation and the ability to attract and use social support are key factors in people's abilities to cope with stress (Palombit, Seyfarth, and Cheney 1997). While social structures are expressed differently across diverse cultures, they remain at the heart of human development and the ability to cope (Mead 1934). Meaningful contacts in family and in group structures, such as those in social team sports, are essential for our mental health.

Supportive social contact is not an absolute prophylactic, but it, along with our predictive abilities, help us combat disease and breakdown (Schulkin 2011). These factors play a direct role in a person's functioning and therefore in his or her potential to respond adaptively in the face of crises. Decreases in extrahypothalamic CRH (for example, in the amygdala) or increased oxytocin as a result of social connection can be recuperative. The individual's cognitive state and social support—or lack thereof—have direct and important effects on his or her ability to cope with adversity and to remain resilient in the face of changing circumstances.

But cortisol can facilitate fear-related events under conditions of adversity, in part by inducting CRH in regions of the brain such as the amygdala. One study on the amygdala and fear response revealed the impact of chromatin modifications, specifically deacetylation of histone 3 at lysine 9 (H3K9), on chronic anxiety-like behavior and nociception (the neural process of encoding and processing noxious stimuli) following a prolonged elevation of CORT in the amygdala (Tran et al. 2014). The molecular processes observed following prolonged CORT activation in the CeA were reversed by treatment with histone deacetylase inhibitors, highlighting the importance of histone deacetylation in the long-term maintenance of anxiety-like behavior and nociception induced by elevated amygdala CORT. The same should hold for the effects of anabolic steroids, used by some athletes, on CRH gene expression.

Growth factors such as neuropeptide Y, which regulate CRH expression, are important in normal regulation. Neurotrophic factor, a kind of growth factor, is also tied to the regulation of CRH in the hypothalamus. Mechanisms for peptides, for instance, include exon duplication and loss, exertion, and gene duplication (Holland and Short 2008). Production of these information molecules is traced back hundreds of millions of years (for example, NPY) (Conlon and Larhammar 2005, Strand 1999). They play a multitude of functions, with roles in food ingestion, metabolic function, and recovery of tissue (Schulkin 1999, Fitzsimons 1998).

Bodily organs are finite systems; they break down. Steroids such as cortisol, which are essential for combating adversity and for sustaining systems, also both inhibit and promote neuropeptides and neurotransmitters. When cortisol levels are elevated for long periods of time, we see breakdown in diverse systems: bone, brain, and immunological capability (McEwen 2007, Sapolsky 1992). One point to remember is that these breakdowns are temporary. For example, during the heat of a basketball or hockey game, some regions may be reduced while other brain regions may expand, as an adaptive response for energy conservation.

The secretion of cortisol, the molecule of energy metabolism, is vital in sport. Cortisol is probably one of the most studied information molecules in the body. This is fitting since it affects everything, including cellular health. One step in battling anxious depression seems to be the activation of cortisol and the regulation of diverse tissue. One much-studied brain region is the hippocampus (McEwen 2007). Adverse conditions that elevate cortisol cause hippocampal deterioration but also recovery. Another result for some people is the aging of neurons and neural deterioration (Sapolsky 1992). Recovery may be greater in some individuals than others.

Under conditions of engagement or adversity and relentlessly high cortisol levels, the amygdala expands as the hippocampus decreases. This is an important feature; the brain is a glucose-eating machine, and its metabolic cost is great. From a neural design point of view, one might expect the quiescence of some regions while others are active because use varies with function and context. Thus, during conditioning or perhaps competitive contact sports, the amygdala expands while the hippocampus reduces (Vyas et al. 2002). This would be an adaptation, but an amygdala that remains enlarged for long periods without regulation and reduction would not be in our best interests. These sorts of events are noted elsewhere in the brain (for example, the striatum), in regions that underlie the organization of action so critical in adaptation in general and sport in particular.

Degradation of hippocampal neuronal functions results from excessive cortisol, leading to a reduction of synaptic density and morphology (Sapolsky 1992). One feature of depression is synaptic degradation and function. Thus, adversity, under some conditions, provokes neuronal deterioration. Drugs such as selective serotonin reuptake inhibitors, or SSRIs, promote neurogenesis in the dentate gyrus (Pinnock and Herbert 2009), but so does physical exercise in animal models.

Exercise generally is important to recovery and resilience. Running, for instance, is known to facilitate neural genesis. Rats and mice placed in a running wheel choose to run, something that Curt Richter discovered in discerning the activity of diverse mammals. Such running promotes the formation of memory and learning (Shors et al. 2001), as we will see in chapter 6.

TELOMERES AND TELOMERASE

A telomere is a region of the chromosome that protects the end of that chromosome from deterioration (Blackburn 2005). Telomerase is a ribonucleic protein enzyme that reduces telomere shortening during chromosome replication and is thus linked to cellular health. A higher level of cortisol (associated with chronic stress and conditions of adversity) has been linked to shortened telomeres (Puterman et al. 2010).

Telomere length is a signal of health. They are, not surprisingly, at their most robust during development, but the length of core DNA is something that reflects overall health across diverse end organ systems and throughout life. Telomere length is also linked to physical activity in a number

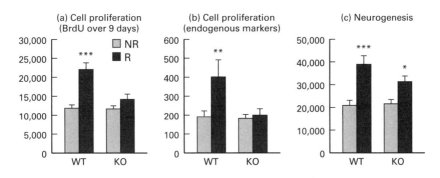

FIGURE 5.6

Impact of short-term running on hippocampal cell proliferation and neurogenesis in Wild Type (WT) and β-endorphin Knockout (KO) mice.

The number of BrdU-IR cells in the whole DG (A), the number of pH3-IR cells (B), and the number of DCX-IR cells (C) were estimated in running (black bars) and nonrunning (gray bars) mice of each genotype sacrificed one day after the last BrdU injection. **P < 0.01, ***P < 0.001 vs. nonrunners.

Source: Adapted from Koehl et al. (2008).

of studies (LaRocca, Seals, and Pierce 2010), being tied to fatigue and to muscle and skeletal strength. Fatigue is associated with short-muscle DNA telomeres; epigenetics suggest that such events occur in strenuous sport and that recovery and regulation are commonplace.

Telomeres are also indicators of cellular health and exercise, physical activity, and sports figure in cellular protection (Puterman et al. 2010). Indeed, an abundance of evidence indicates that sustaining telomere length as we age is aided by hardcore aerobic activities.

CONCLUSION

Comebacks in life and in sport are chancy things: some can do it, some can't, and some can come back to a sport or other activity but then fail. Our emotional sensibility underlies the bulk of activity in sport; our well-being and sense of positive emotion (Fredrickson 2004) figure in our comebacks, sustaining our action and assessment of events. Emotions are active in our adaptive and survival capabilities.

Normal regulation of the internal milieu, much of which is cephalic, underlies all sport and, frankly, all human activity. From regulation to recovery to resilience, common biological mechanisms are operative.

Early social enrichment promotes neural growth factors and neural connections, and sport resilience and diverse practices in sports facilitate the process of adaptation in an interrelated loop of cause and effect. Indeed, the adaptation to fear and other forms of adversity covers sports like a wet, uncomfortable, frosty blanket, punctuated by injury and elation, along with mundane routine and practice and more practice. Of course, with injury and recovery also come vulnerability to much more devolution of function. Sports feature the utter continuity of biology and culture; sport is a wonderful human expression fraught with the very best and the worst in us.

Will we rise from our defeats, or will we be destroyed by them? The Boston Marathon bombings in 2013 killed three people and injured at least 264 others, some of them so severely that they lost limbs. But people rose from the ashes; human spirit rose to the occasion.

FIGURE 5.7

A marathon runner at the 2013 Boston Marathon.

6

RUNNING AND THE BRAIN: NEUROGENESIS

The fossil record indicates that we may have been proficient at walking upright as far back as 4.5 million years ago. Surviving skeletal remains show significant changes in our early ancestors' upper- and lower-limb morphology. When we think of major evolutionary changes that affected the trajectory of our species' development, social contact is clearly one; bipedalism (followed by running) also turns out to be a critical feature in our developmental history (Bramble and Lieberman 2004).

While speed is a factor associated with the purpose of running, it is endurance—the ability just to keep going—that proves crucial to what is perhaps one of the original purposes of running: to follow wounded prey while hunting (Bramble and Lieberman 2004). Running may be the purest expression of the theme of this book: that sport reveals our biological capabilities and that sports enhances these very same functions.

Human history is marked by triumphant acts of sport, and running is one of the earliest documented athletic events. Footraces were a big sport in the ancient Greek Olympics, although they were not marathons per se. Marathons, one of the original modern Olympic Games, are named after the achievement of Pheidippides, a Greek messenger. The traditional story states that he ran from the battlefield at Marathon to Athens to report the Persian defeat, a distance of around twenty-five miles (modern marathons are 26.2 miles), only to collapse and die after shouting "*Nenikekamen!*" (We won!). The ancient historian Herodotos, however, has Pheidippides running from Athens to Sparta (a distance of 150 miles) to ask for help and then running back.

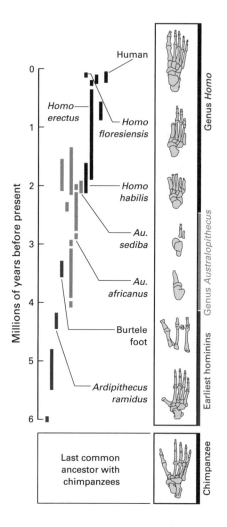

FIGURE 6.1

Walking along the evolutionary tree.

Hominins have evolved many various forms of feet since diverging from the common ancestor they most recently shared with chimpanzees, about six million years ago. The early hominin species *Ardipithecus ramidus* was adapted for both walking and climbing trees, but, like a chimpanzee, had a highly divergent big toe and probably used its feet more like a chimpanzee than like a modern human when it walked. Foot fossils from more recent hominin species, such as *Homo habilis* and *Homo floresiensis*, have a more complete arch. It was probably not until *Homo erectus* that very humanlike feet evolved, with a completely developed arch and a large big toe aligned with the other toes. Feet adapted to both bipedal locomotion and tree-climbing persisted for a long time in human evolution.

Source: Adapted from Lieberman (2012).

FIGURE 6.2

(a) Runners in a race depicted on an ancient Greek vase. (b) Present-day female marathon runner.

Even though Herodotos is sometimes called "the father of lies," his version may be the more accurate one from a biological point of view. While early marathoners in the modern era frequently dropped from exhaustion, marathoning is well within the physical capacities of most reasonably healthy people, with the proper training. Every year, in over five hundred events worldwide, hundreds of thousands of men and women, from young teens to people in their nineties, disabled and able bodied, complete marathons. Elite athletes typically complete marathons in a little over two hours, but the average marathon time is between four and five hours. So popular has marathoning become that multiple marathoning—running a series of marathons—is a growing trend (Mandell 1984, Guttmann 2004).

Clearly, human beings are very good at distance running, something no other animal really does. It is tied to our biological capabilities. Long-distance running evolved from a food-obtaining technique to a necessary social act associated with the spreading of news and from there to one with recreational purposes.

In this chapter, I begin with some of the conditions that set the stage for the act of running and then look at neurogenesis, brain expansion, and longer-term consequences of running within a context of specific

morphological features and diverse information molecules that participate in our capacity for running and sport.

NEURAL PLASTICITY

Seasonal exploration over a vast and ever-expanding terrain for resources
and shelter was necessary for hunter-gatherer societies. The constant
change associated with the hunter-gatherer lifestyle also reflects a division of activity or labor. Indeed, seasonal changes are a common feature
in diverse species; seasonal plasticity in the hippocampus or striatum has
been observed in many seasonal breeders. In fact, birds, with their seasonal
use of song, may have been the first species in which brain changes were
tied to biological adaptation (Marler et al. 1988).

Such plasticity, seasonal and otherwise, is common to cephalic function (McEwen 2007). The hippocampus in particular is perhaps the
best-known and most studied structure in which neurogenesis—the generation of neurons from neural stem cells and progenitor cells—occurs
(Gould 2007). The hippocampus is a region of the brain linked to memory
in a wide variety of functions (Squire 2004). Interest in these regions of
the brain and what biological pathways they were associated with started
in the nineteenth century with classical neuroanatomists who wanted to
describe the hippocampus (Cajal 1906) and understand cellular and fiber
connectivity (Swanson 2003).

Diverse regions of the brain, including the hippocampus, were cited in
debates about what separates us from other primates and in racist doc-
trines about what separates different races. The brain became an object
with which to depict evolutionary ascent, and it was used as evidence in
debates about what is—and what is not—human (for example, Huxley
1863; Owen, Howard, and Binder 2009).

What we do know is that in vertebrates the hippocampus is closely
tied to neurogenesis (Goldman-Rakic 1996), and neurogenesis is tied to
whether one is moving or not. Neurogenesis affects many diverse activities, including the running activity of many species.

For instance, running activity is tied to circadian rhythms and to anticipatory activities that are associated with various resources. Circadian

activities rely on the twenty-four-hour clock, endogenously expressed independent of light conditions, reflecting the heavenly cycles inherent in most but not all living things on this planet.

Importantly, in mammals, running activities are tied to neurogenesis. Plasticity, whether on a daily, weekly, or seasonal basis, is a key feature of the mammalian brain. Diverse cyclical clocks regulate the brains of mammals.

Curt Richter used the running wheel or activity cages with rats to depict the rhythmic features of behavior that could be disassociated from external signaling systems. In his studies, circadian rhythmic cortisol was shown to be linked to cephalic circadian rhythmic activity. Cortisol is often viewed as being associated with the "wake-up signal," in addition to its many other important regulatory purposes.

Mammals are not alone in being regulated by circadian cycles; clock-like neural circuits are found even in invertebrates. Circadian-like behaviors are tied to the anticipatory actions of flies and other insects. One neuropeptide tied to anticipatory behaviors in both vertebrates and invertebrates is corticotropin-releasing hormone (CRH), a particularly ancient neurotransmitter.

Studies using mammals have shown us that running in a running wheel is strongly linked to hippocampus neurogenesis and that diverse peptides, such as brain-derived neurotrophic factor (BDNF), are linked to both running and neurogenesis and recovery. BDNF is one of the many peptides that are tied to sustaining tissue fundamental in adaptation and sport and to regions of the brain such as the hippocampus.

Studies that used mammals to examine running's effect on the brain noted that the test subject would continue to run whether or not it received an external reward for its efforts. It seemed instead that the very act of running was providing positive reinforcement and that no additional reward was necessary to encourage the continuation of the behavior. Perhaps this offers a clue as to why running became such a profound feature of sport and continues to be a primary exercise routine for many human beings.

Seasonal breeding and seasonal exploration, in addition to core learning, are examples of plasticity in brain activation and changes in neural function as a result of context and circumstances. Changes in neural expression (gray matter, connectivity) are now noted in diverse contexts in sports-related events (Babiloni et al. 2010). These events are both genetic and epigenetic—in other words, the events are under the influence of both genetics and experience. Indeed, the events are continuous with one another in everyday life. Moreover, dihydrotestosterone, among other anabolic steroids, underlies the epigenetic neurogenesis that occurs with exercise.

Learning and memory, which are both required for successful participation in sport, reflect practice and skill as well as background capability (some of us are much more capable than others at this). But all humans have some running ability and will exhibit the same sorts of responses to a running regimen.

Brain expansion and then decrease is an energy-conservation feature of neural design; parts of the amygdala are tied to appraisal of unfamiliar events. Uncertainty is tied to amygdala expansion and hippocampal shrinkage. The hippocampus region is linked to memory in particular (Eichenbaum and Cohen 2001). The movement of hunter-gatherers probably reflects such functional relationships (for example, remembering where resources can be found in hunting and foraging for food).

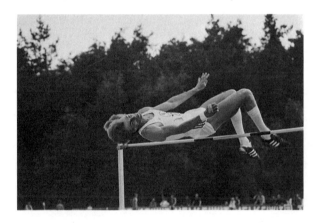

FIGURE 6.3.A

Dutch Athletics Championships 1979, Nijmegen; the high jumper Ruud Wielart in action.

FIGURE 6.3.B

Southern Counties twelve-stage road relay race, Wimbledon Common, London, 1988.

FIGURE 6.3.C

Runners in a relay race, Brisbane, 1939. Val Weaver and Vera Askew passing the baton.

Source: John Oxley Library, State Library of Queensland.

Diverse information molecules are tied to neural plasticity, and all of them can in the proper context act as growth factors. Researchers uncovered a common chemical messenger in growth factors (a diverse family of peptides) that underlie developmental trajectories; the Nobel Prize for uncovering this went to Rita Levi-Montalcini and Stanley Cohen in 1986. Some of these growth factors include activin, colony-stimulating Factor, connective-tissue growth factor, epidermal growth factor, erythropoietin, fibroblast-growth factor, galectin, growth hormone, hepatocyte growth factor, insulin-like growth factor binding protein, insulin, insulin-like growth factor, keratinocyte growth factor, and leptin.

MORPHOLOGY

Many paleoanthropologists now suspect that long-distance running may be a specific evolutionary adaptation to group hunting over long distances that evolved specifically with us (Lieberman 2011). We are socially cohesive predators, and lower-limb physical adaptations may have emerged about two million years ago to facilitate long-term trotting while tracking prey. Other related adaptations include:

- Enlarged gluteus
- Small waist
- Neck thorax flexibility
- Expanded flexibility of vestibular and ocular reflexes

Calcaneus length is tied to this evolutionary trend, facilitating a morphological design that favors efficiency in long-distance running. The length and flexibility of the Achilles tendon, for instance, is critical in warm climates for distance runners (Raichlen et al. 2013).

The diverse skeletal features depicted in table 6.1 are physical characteristics that make long-distance running possible in our species (Bramble and Lieberman 2004). Diversification of the foot played a key role in the origins of locomotion and running. By the time *Homo erectus* emerged some 3.5 million years ago, a modern foot, almost indistinguishable from ours, had evolved (Lieberman 2007).

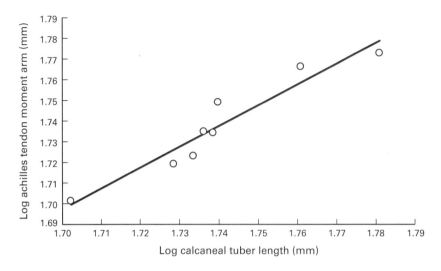

FIGURE 6.4

Relationship between Calcaneal Tuber Length and Achilles tendon moment arm measurements.

The correlation is significant (r = 0.95; p = 0.0002). The following ordinary least-squares regression line may be used to determine actual moment arm lengths from isolated calcanei (slope [95% CI] = 1.00 [0.31]; intercept [95% CI] = −0.01 [0.54]; r² = 0.91).

Source: Adapted from Raichlen, Armstrong, and Lieberman (2011).

TABLE 6.1 **Derived features of the human skeleton with cursorial functions**

FEATURE	FUNCTIONAL ROLE	EARLIEST EVIDENCE
More balanced head, short snout	Head stabilization	*H. habilis*
Tall, narrow body form	Thermoregulation	*H. erectus*
Forearm shortening, narrow thorax	Counterrotation of trunk	*H. erectus*
Narrow pelvis	Counterrotation of trunk versus hips	*Homo?*
Stabilized sacroiliac joint	Trunk stabilization	*H. erectus*
Long legs	Stride length	*H. erectus*
Shorter femoral neck	Stress reduction	*H. sapiens*
Long Achilles tendon, plantar arch	Energy storage	*Homo?*
Permanently adducted hallux	Stability during plantar flexion	*H. habilis*
Short toes	Stability during plantar flexion Distal mass reduction	*H. habilis*

Source: Adapted from Bramble and Lieberman (2004).

Key morphological features of *Homo erectus* are its short toe and long Achilles tendon (Lieberman et al. 2013). These features contributed to endurance capability in exploring and hunting and are associated with speed. Some primates and apes (for example, chimpanzees) are bipedal for short bouts and in some contexts, but their hip, spinal, and limb structures do not make being bipedal an optimal mode of locomotion for long periods of time, and they are certainly not efficient runners. Bipedalism, which took shape as an experiment of nature over the last five million years, may have also aided movement in trees (Thorpe, Holder, and Crompton 2007).

The subtle morphological changes that made it possible for *Homo erectus* to move bipedally also allowed it to enlarge its territory, which may be tied to an expanded brain and increased technological capability. One result of bigger brains is, of course, greater cognitive/affective capability, resulting in exploration and the development of technology and culture (and eventually sport).

Efficient bipedalism requires a narrower pelvis than that found in australopithecines. Bigger brains and a narrower pelvis, however, means a difficult parturition. *Homo erectus* females probably would have required assistance from others (as do modern women) in giving birth, and their infants were likely born at an earlier stage of neural development than those of other primates. This has implications for hominin social structure, indicating a certain level of cooperative social behavior and an extended juvenile stage. Diverse hormonal developments also may be involved in human birth and parturition patterns. The nuclear progesterone receptor (NPR) gene might be involved in this evolutionary process, linking bigger brains and the duration of human pregnancy and parturition.

Running speed is, of course, tied to the length of the leg and the stride, within a context of conservation of energy and maximization of resources through the utilization of glucose and the maintenance of fluid volume and loss. Steroid hormones such aldosterone, an adrenal steroid hormone, are tied to fluid volume and sodium conservation, which are essential in maintaining fluid levels, solute volume, and tonicity for continued viability (Denton 1982). Thus conservation at all levels is operative: sodium and water excretion drop under conditions of long-distance running in the tropical climate in which humans evolved.

Perhaps our need to roam far is related to our nutrition needs. We were and are meat eaters, although we are more accurately categorized as omnivores (Rozin 1998). We can manage as vegetarians or even vegans—hundreds of thousands of human beings survive and even thrive on such diets. We can do quite well on a meat-only diet; Eskimo and Inuit have done so for thousands of years. But the range of nutrients and fiber we require is most easily satisfied by a combined meat-and-plant diet. The range of what hominins from *Homo erectus* on ate, and most definitely what we eat, reflects omnivory.

Perhaps 3.5 million years ago, the feet of related hominoids indicate they may have both climbed trees and walked erect, with long toes, a grasping big toe, and a flexible arch. Bipedalism with such a foot would have been effective, but these hominins were likely not able to run distances. Changes in the foot, as our hominin ancestors committed more and more to a ground-dwelling existence, set the conditions for our capabilities associated with running, coupled with the expansion of the cortex, which facilitated an increase in all our physical capabilities.

Bipedalism is one very important adaptation, and a humanlike gait can be traced back to *Australopithecus africanus* some four million years ago. In addition, the expansion of the shoulder muscle, the Achilles tendon, and the larynx (for speech and singing) (Lieberman 1984) are adaptations that set the stage for our social and physical engagement with sports.

METABOLISM

Our metabolism evolved for longer-term movement while treading over great distances, and our explorations often required endurance while navigating through environments where excessive heat loss was a concern. Our ability to regulate our internal milieu to account for variable environmental conditions is yet another evolutionary adaptation that helped our ancestors continue to traverse great distances and harsh environments.

The brain is an active organ. A massive absorption of glucose is required to sustain its energy expenditure, the activity of the neuronal assemblage that underlies its diverse activities. Indeed, glucose utilization is a core feature of metabolic regulation in every organ system but particularly the brain; glucose utilization is a marker of brain activity and can be used to mark relationships, for instance, between memory and learning. The technique for measuring 2-deoxycluse utilization resulted in a Nobel Prize for Julius Axelrod.

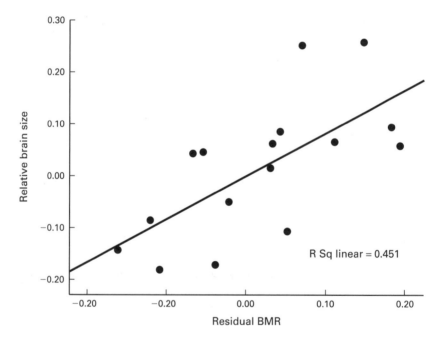

FIGURE 6.5

Basal metabolic rate (BMR), controlling for body size (residuals from a linear regression of basal metabolic rate versus log-transformed body size), plotted against relative brain size. Species with higher metabolic rates than expected for their body size also have larger-than-expected brain size. Higher metabolic rates are tied to (for their body weight) higher-than-predicted brain size.

Source: Adapted from Dunbar and Shultz (2007).

Glucose utilization maintains functional architecture in the brain's networks that underlie skill in sports or memory and learning. Brain activation may account for roughly 70 percent of energy demands in the brain (Dunbar and Shultz 2007).

The metabolic costs of walking are similar across higher primates and therefore early hominins (Pontzer, Raichlen, and Sockol 2009), but walking became less metabolically expansive with greater motor dexterity. Energy expenditure is a key evolutionary feature, and walking upright is tied to our evolution because it allowed us to see better and maximize our resources. Standing and looking efficiently are part of our evolution. Indeed, standing straight may have been a metabolic advance (McHenry 1994).Walking, perhaps by the middle of the Pliocene for hominins, was more efficient and less metabolically demanding than the ground locomotion of other kinds of apes (Pontzer, Raichlen, and Sockol 2009).

When one compares the efficiency of locomotor expenditure (Pontzer, Raichlen, and Sockol 2009), selection pressure favored short hind legs and long arms in chimpanzees because that sort of structure is more efficient than standing upright for getting up and moving around in trees (Pontzner and Wrangham 2004).

Metabolic rates and regulation of running and walking are key features of our species' survival since our ancestors moved from one area to another. Sheer physical capability and brain size are correlated (Raichlen and Gordon 2011), and metabolic rate serves as an indicator of the regulation events essential toward this end.

In other words, the evolution of the brain is tied to physical capability, running being one feature of this evolution. Indeed, metabolic rate is a feature of athleticism, as are brain mass and physical capability. One of the more interesting findings in recent years, however, is the correlation between physical or exercise capacity and the resulting brain size in diverse mammals (Raichlen and Gordon 2011).

Diverse studies have demonstrated the links between brain activation and glucose metabolism in the brain and other end organ systems. Thus the size of the brain region and the expanded utilization of glucose under normal conditions facilitate an expansion in energy expenditure and allows for running.

TABLE 6.2 Brain mass and MMR data

MAMMAL	BODY MASS (G)	BRAIN MASS (G)
Alopex lagopus	4,510	35.52
Antilocapra americana	28,400	145.78
Apodemus sylvaticus	20	0.58
Bettongia penicillata	1,100	9.56
Bos taurus	475,000	454.40
Canis familiaris	25,900	79.99
Canis latrans	12,400	83.37
Canis lupus	27,600	132.75
Capra hircus	24,300	110.50
Cavia porcellus	584	4.51
Connochaetes taurinus	102,000	364.33
Equus caballus	453,000	702.50
Gazella granti	10,100	148.74
Genetta tigrina	1,380	15.18
Helogale parvula	430	4.76
Homo sapiens	77,940	1311.40
Kobus defassa	110,000	314.61
Madoqua kirkii	4,200	34.31
Mungos mungo	1,140	10.49
Mus musculus	26	0.44
Neotragus moschatus	3,300	33.18
Ovis aries	21,800	132.50
Panthera leo	30,000	238.50
Peromyscus maniculatus	22	0.63
Rattus norvegicus	278	2.31
Spalax ehrenbergi	136	1.88
Sus scrofa	18,500	186.60
Tamias striatus	90	2.20
Taurotragus oryx	240,000	460.00

The majority of brain mass data were compiled by Isler and van Schaik (2006). All MMR data were compiled by Weibel et al. (2004).

Source: Adapted from Raichlen and Gordon (2011).

ENDORPHINS AND OTHER INFORMATION MOLECULES

A major discovery during the age in which many peptide hormones were discovered was that of the endorphin hormones. The possibility that chemical signaling included messages that reduced pain or were tied to the highs associated with being on drugs was theorized and then confirmed.

The endorphins consist of a large group of peptides that date back at least a half-billion years. They are expressed in both vertebrates and invertebrates, in which they are diversified in expression through various end organ systems (for example, the brain, the pituitary gland) and in concert with other peptide hormones. These peptides, like many others, were discovered in the brain during the great revolution in biochemistry during the 1970s and 1980s.

It is still not completely clear what the roles of these diverse endorphins are, but they range from the reduction of pain to euphoria and to positive links associated with social attachment (Koob and LeMoal 2005, 2008). One of course wants to distinguish diverse endorphins expressed in the central nervous system from the peptides expressed in the brain itself and those expressed in the periphery and to distinguish the peptides detected in extracellular fluids from those detected in the cerebrospinal fluid. A blood-brain barrier keeps peptides in the periphery.

Endorphins in the central nervous system are site specific in terms of whether they enhance pleasure, and it has been found that endorphin injections induce pleasure. Moreover, endorphin-like substances in the periphery are either elevated or reduced during drug withdrawal and not necessarily elevated during drug ingestion or pleasure. But most forms of psychoactive substances result in the elevation of endorphin-like substances.

One important neurotransmitter is dopamine. The interaction of dopamine and endorphin-like substances may underlie part of the good experience that is tied to long-distance running: the neurotransmitter, perhaps, for long-term stability and the neuropeptides for fleeting moments of euphoria.

Here is how Yiannis Kouros, an ultrarunner (one who runs greater-than-marathon distances), described his euphoria:

> Some may ask why I am running such long distances. There are reasons. During the ultras I come to a point where my body is almost dead. My mind has to take leadership. When it is very hard there is a war going on between the body and the mind. If my body wins, I will have to give up; if my mind wins, I will continue. At that time I feel that I stay outside of my body. It is as if I see my body in front of me; my mind commands and my body follows. This is a very special feeling, which I like very much. . . . It is a very beautiful feeling and the only time I experience my personality separate from my body, as two different things.
>
> (http://www.lehigh.edu/~dmd1/sarah.html)

Euphoria is reported in many studies of running. Running, as with other forms of physical exercise, can be viewed as a way of reducing stress-related events. Diverse studies over a number of years have found endorphin levels in runners to be elevated. Use of PET to measure endogenous endorphins both before and after a two-hour run found dramatic changes in the expression of endogenous endorphins in the binding of the radioactive ligand (Boecker et al. 2008). There are visible changes in the opoidergic binding in neurons in the frontal areas, limbic/paralimbic areas, and temporoporietal areas of the brain when there are changes in the Visual Analog Scale (VAS) ratings of euphoria (Boecker et al. 2008).

Of course, it's not all running highs and happy feelings. Long-distance running also involves combating pain and discomfort—common themes in many sports. Adversity is essential to sport. To struggle is to succeed, and to cope with struggling, the human body has evolved to release hormones associated with euphoric states. When one is faced with a particularly trying physical feat, the cephalic space is permeated with a sense of calmness, and what seemed daunting ceases to be much of a bother at all.

This would implicate both central and peripheral physiological signaling systems to induce states of quiescence, of "no worries." Some runners do experience periods of sheer intensity, but a sense of having no worries predominates. Another theme is a sense of accomplishment. Having achieved a perceived goal is one feature of a runner's high. During sports of any kind, the body is being pushed to its limit and often beyond it, but the euphoria that comes with great physical exertion delivers the reward needed for continued physical pursuit regardless of the exhaustion and pain otherwise associated with that activity.

Pleasure is not one thing, after all, but a grab bag of states; so are the emotional states derived from diverse sports. Various forms of effort and relief are thus linked to different information molecules acting in diverse ways in shared brain regions.

Regions of the brain—including the nucleus accumbens, a region of the basal ganglia tied to the organization of movement—are responsive to information signals, including those of endorphins or opioids (Peciña, Smith, and Berridge 2006). For instance, endorphins infused into the frontal end of the nucleus accumbens elicit liking and inhibit the rejection of food sources. Infusions of endorphins into this region apparently enhance the sense of sweetness in food (Peciña, Smith, and Berridge 2006). There are likely more caudal regions of the brain linked to the rejection of substances. My colleague Kent Berridge calls this region part of the hedonic circuitry that underlies behavior. This region has long been thought of as the link between motivation and the motor system (Swanson 2000).

But liking something is one thing; the motivation to train for it is quite another. Ask any athlete; dopamine is clearly tied to desire (Berridge and Robinson 1998) and, among other things, effort. The same information molecule infused and active in the amygdala enhances the salience of events (Mahler and Berridge 2009). The interactions of these two regions, the nucleus accumbens and the central nucleus of the amygdala, represent a link between motor systems and motivation via a common type of information molecule: the endorphins. The endorphins are tied to the liking, the euphoric sense of achievement of finishing a marathon, pain and all.

ENDOCANNABINOIDS

The endogenous endocannabinoid is tied to the stress response, the sense of adversity that underlies human activity in general and sport in particular. The reduction of endocannabinoid signaling is an adaptation to adversity, as evidenced by the management of the HPA axis (Hill et al. 2010). The breakdown of this signaling system contributes to negative states. Mice genetically modified to lack this endocannabinoid ran half as much as they would normally—and mice normally run a lot.

One view of the endocannabinoids is that they are protective under conditions of adversity (Hill et al. 2010). Long-distance running is but one example of information molecules playing protective roles in long-term tissue viability.

Exercise-induced elevated levels of endocannabinoid (eCBs) signaling occur in humans and several other species (Raichlen et al. 2013). In humans, treadmill running and intensity are linked to endocannabinoid signaling. The idea is that eCBs is tied to reinforcing properties of "runner's high" (Raichlen et al. 2013). We also know that a variety of other information molecules are altered in the brain via running behavior (neurotrophic factors, for instance). Interestingly, exercise capability is linked to a feature of cortex expansion (Raichlen and Gordon 2011).

Cannabinoid receptors such as anandamide (AEA) are linked to reward signaling, but that description of purpose is probably too narrow. What we do know is that these information signals are expressed and regulated during sports such as running. In comparative studies in humans, dogs, and ferrets, cannabinoid receptors were more elevated after running than after walking (Raichlen et al. 2013).

For instance, humans who had an anandamide concentration of 2.2 pmol ml before running had that concentration jump to 6.1 pmol ml after running for thirty minutes at a Froude number of 0.70, which was calculated by $Fr = velocity^2/(gravitational\ acceleration \times length)$. After walking for thirty minutes at a Froude number of 0.25, humans who started with an anandamide concentration of 1.7 pmol ml saw their anandamide concentration sink to 0.9 pmol ml (Raichlen 2013).

Exercise affects diverse regions of the brain. Aerobic exercise increases anterior hippocampus size (Erickson et al. 2011, Chaddock et al. 2010). The studies above are also linked to the improvement of memory, which reflects animal studies on the improvement of learning as a function of running activity. Higher fitness is associated with greater expansion not only of the hippocampus but of several other brain regions. Importantly, children who participate in aerobic activity or fitness have enhanced

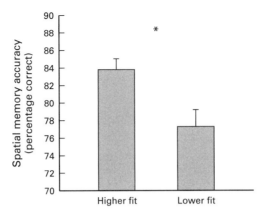

FIGURE 6.6

Children classified as higher fit showed higher accuracy across all dot conditions of the spatial memory task one year later. Error bars represent standard error. *p < .05.

Source: Adapted from Chaddock et al. (2011).

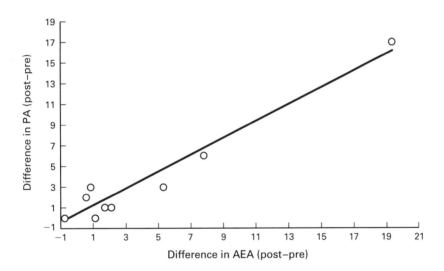

FIGURE 6.7

Correlation between positive affect (PA) and AEA in humans.

Values are the difference between pre- and postexercise PA scores plotted against the difference between pre- and postexercise plasma levels of AEA. Note that the values for two subjects were nearly identical (difference in AEA = 0.89 and 0.90, difference in PA = 3 for both subjects), and they are not differentiated on the figure.

Source: Adapted from Raichlen et al. (2013).

cognitive capabilities (Chaddock et al. 2011). Basal ganglia volume is associated with more fit adolescents (Chaddock et al. 2010). In addition, aerobic fitness enhances cognitive function and frontal cortex lateralization of function in older males (Hyodo et al. 2016).

Perhaps such events occur over the complete life cycle. Aerobic capability or fitness is also associated with hippocampal volume in older individuals (between sixty and seventy years old) and their ability to perform memory tasks (Erickson et al. 2009; Chaddock et al 2010). Thus, increases in volume expansion in both younger and older adults are associated with aerobic exercise. Clearly, aerobic activity is beneficial both in childhood and for older people, and brain expansion and memory capability are tied to aerobic activity levels (Chaddock et al. 2011). Indeed, plasticity in the brain as a function of exercise is quite remarkable. For instance, the bilateral hippocampal volume in groups with lower levels of fitness was around 6,850 mm^2 but around 7,800 mm^2 in groups with a high level of fitness (Chaddock et al. 2010).

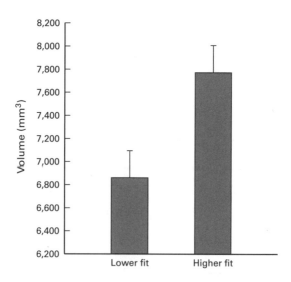

FIGURE 6.8

Bilateral hippocampal volume as a function of aerobic fitness group. Error bars represent standard error.

Source: Adapted from Chaddock et al. (2010).

CONCLUSION

Running is an expression of our evolution, and, amazingly, running capability is linked to neural expansion. So not only are social and linguistic capabilities tied to an evolved cortex, but so is running capability. Diverse physiological signals facilitate adaptation and increased performance in sports. Of course, speed and size interact with shape and endurance capabilities.

Neurogenesis is a feature of several regions of the brain, most notably the hippocampus. Running-induced activity facilitates the expansion of the hippocampus in a number of species. Running itself promotes cell proliferation in the hippocampus, in part through the induction of endorphins or diverse neuronal growth factors (Koehl et al. 2008). Running and neurogenesis are linked to forms of basic adaptation; running easily transitioned from joint coordination to play and eventually to sport.

Diverse information molecules (cortisol, endocannabinoids, dopamine) are essential in athletic activity and in activity in general. Growth hormone or endorphins (Koehl et al. 2008) facilitate and sustain the expression of neurogenesis in regions of the brain such as the hippocampus in paradigms such as running activity.

Athletic capability, effort and exercise, and sport and physical play reveal a positive impact on neural function. In athletics specifically and in life more generally, striving, desiring, and succeeding are constants. The metaphors surrounding the adrenal rush (wanting) and the endorphin high of being satisfied run across the great array of human experience in general and sport in particular.

What matters is the clear link from walking to running to sport within a context of appraisal systems. In general, and in many sports, we are chronically appraising events, such as in our ability to infer strength from sounds. The athlete battles toward his or her development of efficiency, excellence, and reward—and, depending on the sport, pain and drudgery—but we will get to that later. Here, using the paradigmatic example of running, we see that the biological and evolutionary thread connecting our primitive behavior to the social context of sports is easily traced.

7

THROWING, SWIMMING, AND ROWING

Human evolution has not been a simple matter of physical change. As our bodies evolved and our brains expanded, humans developed significant behavioral patterns—ways of managing and adapting to the world. These behaviors are crucial in all human activity, and they exercise broad influence over sport.

A strong human motivation is to explore—to traverse the unfamiliar while taking along the familiar (Rozin 1976). We do this with foods, for instance: think of familiar sauces poured onto new foods. Always a drive for something new, but mixed with the safety of something known. Sport is rich in the mixture of the familiar with the less familiar; in sport we apply our habitual patterns of exercise and performance to new goals and physical challenges. The regularity grounds the action, but the action is rich and novel in anticipation, imagination, and expression.

In this chapter we continue our theme of how we anchor to objects and to others in the organization of action and in sport in particular, focusing on social organization, human evolutionary development, pedagogy, and tool use. I start with the visual system, which is central to human evolution and our practice of most sports. I will then continue with problem solving and sport, throwing, swimming, and rowing.

VISION

Even before we became bipedal walkers, long before we came down out of the trees, early primates evolved stereoscopic vision. It is their gaze that lends even the most primitive modern primates an eerily human appearance. Visual contact and stimulus is integral to our species' early infant and childhood development, as it allows us to learn from one another—to learn about forming necessary social bonds and to understand how to act and survive in one's environment. Vision is dominant in our evolutionary progression, noted by its dominance across the neocortex; indeed, the greater an animal's sociality, the larger its visual cortex tends to be. Good vision, along with an awareness of visual context, allows for success in play, followed by games, and eventually by competitive sport (Barton 2006).

A modern map of the cortex places vision as the most dominant means of cortical expression. Indeed, we are endlessly visual, and vision features prominently in almost everything we do. Although blind people manage perfectly well in a modern culture and often exhibit compensatory expansion of auditory function in cortical expression, blindness would have been a devastating and likely fatal disability in the prehistoric world. Comparing human brains with those of other animals, the degree of visual expansion, both cortical and noncortical, is apparent. Visual cortical and motor cortical connectivity is impressive in the extent of neural connectivity to other cortical and subcortical regions of the brain.

We know that our social capabilities are highly correlated with the expansion of cortical tissue (Barton 2004, 2006), a common theme in primate evolution. This is particularly true with the visual cortex and the cortex more generally.

In mammals with forward-facing eyes, such as most carnivores and the primates, retinal axons are routed so that visual information from the same points in space coming from the two eyes can be combined. From both eyes, ganglion cells whose receptive fields are in one half of the visual field project to the opposite cerebral hemisphere. This cross-routing occurs at the optic chiasm. Here, axons from the medial (nasal, or nearer the nose) half of one retina cross over to join the axons from the lateral (temporal, or nearer the temples) half of the other retina. In other words, the temporal retina projects to the *ipsilateral* (same-side) hemisphere and the nasal retina projects to the *contralateral* (opposite-side) hemisphere. Unlike somatic sensation, which is entirely crossed, only half of the retinal axons cross. Like somatic sensation, however, the crossing of retinal axons results in both halves of the external visual field being represented in each cerebral hemisphere.

We are a species oriented to looking at objects, at others, and at what others are looking at, as well as to looking at objects together. Looking at each other is reflected in the expanded volume of the primary visual cortex (Dunbar and Shultz 2007) in comparison with that of several closely related species. The expansion of the visual system is most palpable in terms of the visual cortex (Barton 2004). The degree of visual cortex and binocularity are fundamental features of visual expansion in our species when compared with other species (Barton 2004, 2006; Van Essen 2005; Van Essen, Anderson, and Felleman 1992).

All aspects of our coordinated adaptive systems are linked to visual function. For instance, visual systems guide motor control. Visual expansion is perhaps reflective across the distributed network of the visual system, which includes both neocortical and other regions (Barton 2006). Standing upright and the expansion of diverse muscular capabilities (for example, the shoulder muscle and Achilles tendon) is tied to binocular visual acuity (Barton 2004).

Our specific visual adaptations are critical in our sporting activities. Consider the visual excellence required in archery, especially the specialized skill of archery on horseback. The rider must simultaneously process and react to the action of the horse, potential obstacles in the horse's path, the changing angle and position of the archery target as the horse moves forward, the placement and aim of the bow, and the projected path of the arrow.

While visual acuity developed because of its advantages in foraging, feeding, and hunting, we now use our cephalic visual capability in sport. Aside from archery, pitching and golf make enormous use of our sense of geometry, a basic visual and cephalic capability that is innate, not learned. What is learned is the *instrumental* use of geometrical capability. But we arrive prepared to examine and understand objects.

OBJECT KNOWLEDGE

Understanding objects is, like most forms of knowledge acquisition, a kind of contact sport, especially for the very young: grasping, engaging, manipulating, etc. Within this early period of life our memory develops (Kagan 2002). It is not so much that children change their orientation to objects as they grow (Carey 1985), even though their knowledge base certainly changes and expands. It is just that sensorimotor exploration is already embedded in theory, in notions of objects, in considerations of what is alive or not, concepts of what animals might be or not.

Children categorize objects (manmade or natural, animate versus inanimate) (Keil 1989) at a very young age. Some developmental cognitive experts see children as little scientists (Gopnik and Meltzoff 1993), and certainly that is consistent with the idea that science emerges from common problem solving, from a set of cephalic capabilities, search mechanisms, sampling, tracing, and tagging events that matter. Indeed, keeping track of what we are foraging for is a key feature in our evolutionary and ontogenetic development.

Orientation to end organ systems is a basic feature that children have about object knowledge and its importance in the organization of action and decision. The lexicon expands, but the orientation is still quite basic (Carey 1985, Keil 1979, Spelke 1990). Knowledge of the brain, for instance, is just a further expression of a basic capability and an adaptation in the organization of knowledge, tagging, and tracking events.

Young children come prepared to link kinds of objects together into category-specific features. Indeed, children as young as five infer the underlying properties of the brain in explaining human behavior, intentions, and other properties (Kagan 1984). And of course, all of this figures in sport.

Most of these events are linked to visual object knowledge; other sensory features involved are often linked to basic nutritional issues (salt and sweet taste, for instance, is tied to metabolic needs). Object identification in action is a key feature in getting anchored to real-world objects.

At birth we have a toolbox of concepts for cataloguing and tracking objects across the visual cortex, for discerning meaning as to what something is and where it is. These skills are instantiated in diverse neural systems that represent the sensory features of the objects (Ungerleider and Mishkin 1982). We are particularly adept at object knowledge about the human body. Accurate computerized facial recognition is still a bit of a pipe dream, but most humans can pick individual faces out of a crowd with great speed and rapidly discern motive and emotional state by observing gaze, expression, and body attitude (Dunbar 2010)—all of which are critical in a variety of sports.

The expansion of vision, our upright stance, and the use of tools are evolutionary factors closely linked to object knowledge and perhaps tied to Bayesian reasoning in rating prior probabilities and self-correction, a skill vital for survival and for prediction in general. Object knowledge is also critical in sport. Consider what it takes to pick out the relevant person in a social sport such as basketball, hockey, or lacrosse and get the ball to them. Face and body are coded in these events, and the direction of their eyes and bodily attitude all affect where the ball is delivered—and precise, excellent, and rule-conforming ball delivery is a normative goal in many sports.

PROBLEM SOLVING: SPORT

Sport is rich in problem solving.

One cephalic tool for exploring these links is a Bayesian capability to check on hypotheses and modify our frameworks and codified habits via feedback. In other words, no one unifying super-problem-solving device exists; rather, we have a wide assortment of adaptive tools that underlie problem solving, involving the evolution of adaptive systems. But cephalic function and adaptive systems may inherently entail probabilistic expectations of diverse forms of sensory information and ways in which to track, link, and structure events in meaningful categorical relationships that underlie causal inferences (Goodman et al. 2001).

The movie *Moneyball* highlights the importance of the use of statistical scanning in picking possible successful players in baseball by tracking how individuals function in a collective context. This is now being used in basketball and hockey teams. "Analytics" is a management position that has opened up in recent years across many sports. Most baseball teams, and now many basketball and hockey teams, employ an analytics expert. One example: John Hollinger (2002) made a career as a statistician for ESPN extracting basketball data; he developed the Player Efficiency Rating (PER), which is now widely used. In December 2012 he was recruited to be director of operations for the Memphis Grizzlies to help put together a roster. Thus, statistical scanning is used in basketball and no doubt will spread across diverse recruitment strategies for professional sports.

The sport statisticians use, among other analytical and statistical methods, Bayes' theorem (a fundamental theorem of probability that states that, for any two events A and B, the probability of A given B can be computed from the probability of B given A, as well as the overall probabilities— known as the "prior probabilities"—of A and B). This is one way to understand expectations in terms of prior probabilities that are embedded in our predictions about recurrent events. As Clark notes (2013) insightfully, these events lie within a wide array of problem-solving capabilities, an array without one set of consistent overarching rules (Rozin 1976).

$$p(A \mid B) = \frac{p(B \mid A)\, p(A)}{p(B)}$$

Moreover, we may be prepared to understand natural frequencies more easily than other ways of representing events as we track objects and events with diverse heuristic devices and revise our orientations to events if we need to (Gigerenzer 2000).

A Bayesian cephalic self-corrective functions to coordinate expectations with new forms of evidence in a context of predictive capabilities. This is perhaps one part of our cephalic capabilities—a grab bag of diverse functions (Rozin 1976) that underlies human foraging for coherence and sampling behaviors. One of the nice things about sports is that it breaks down conceptual barriers that we have constructed, such as barriers between theory and practice, perception and action, and among thought, action, and self-correction. In Figure 7.3, Agiloti et al (2008) demonstrate how skill is related to a player's and novice's predictive capability.

Nate Silver's book *The Signal and the Noise* (2012), though well known for popularizing Bayes' theorem and for predicting outcomes in sport and politics, is oversold; there are other statistical analytic tools (Hacking 1964, 1975, 1999). But Silver brought into a public context something we do all the time: statistical analysis. A core feature in our evolution is predictive capability and the use of prior experience in self-corrective contexts, and the context in sports and many other areas of life is action oriented. Players (of basketball, football, soccer, lacrosse, etc.) determine in milliseconds the prior probabilities of their teammates' and opponents' movements: statistical inference, space, time, and probability are effortlessly built into cephalic capabilities.

Prior motor learning and experience, not surprisingly, affect athletes' learning of movements and their physical capabilities (Shadmehr and Krakauer 2008). Indeed, motor learning, endemic to diverse sports, is tied to prior learning and action (Yarrow, Brown, and Krakauer 2009). Prior probabilities help determine what our next motor sequences will be. Sport is fast, frugal, and adaptive, and these are core features of evolution and adaptation. Tracking probabilities in action underlies our evolutionary success; it underlies sport in particular and human action in general. Bayesian-like inferences are one mechanism operative in cephalic capability in action. Information trajectories are embedded in the organization of action and the prediction of oncoming events (O'Doherty et al. 2004).

One model of motor control and of sport is provided by Yarrow, Brown, and Krakauer (2009). It consists of an anticipatory control of cephalic regulation of movement in general and sport in particular, which is how we understand brain/behavior relationships, and in which diverse information molecules (for example, dopamine) underlie the various features of training and competing. What runs through this model is anticipatory motor control, which is endemic in the organization of action (Dewey 1895, 1896). Feedforward anticipation and then corrective cephalic capability are critical in the organization of action (Yarrow, Brown, and Krakauer 2009).

Furthermore, striatal dopamine facilitates the interaction between cognition and movement, which are, as discussed earlier, significant contributions to problem solving in sport (Rauch, Schonbachler, and Noakes 2013). Throughout our evolution, our learning capabilities have become enhanced in regards to sensorimotor control. This evolved sensorimotor control is embedded in cognition, for which dopamine is a primary neurotransmitter. The neural networks that stretch from the forebrain to the brainstem have evolved over time to become nontrivial in our reaching and grasping of objects in general but also in sport in particular (Begliomini et al. 2014, Alstermark and Isa 2012). And reaching for and grasping objects also figure across sports with respect to anticipatory behaviors.

When it comes to problem solving, then, the brains of elite athletes are different from other people's—but not necessarily in obvious ways,

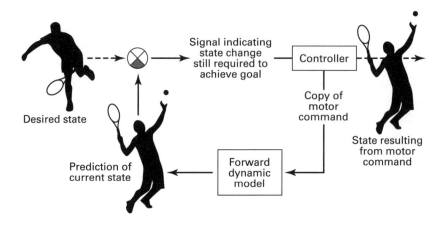

FIGURE 7.1

A key idea in computational motor control is that the brain, through an internal stimulation known as a "forward model," is able to predict the imminent change in the state of either a body part or an object that will result from an outgoing command. For example, when you move your hand from one place to another, the brain can estimate its new position before sensory feedback arrives. An optimal estimate of your hand's position can be obtained by integrating the forward model's prediction with actual visual and proprioceptive feedback. Forward models can also be trained. When discrepancies arise between sensory feedback and a forward model's prediction—for example, when one is wearing prism glasses—the forward model can adapt to reduce the prediction error.

Source: Adapted from Yarrow, Brown, and Krakauer (2009).

any more than the brain of Einstein is obviously different from ours. The genius of the elite athlete is how diverse cephalic capabilities are put together in a uniquely excellent form: this excellence derives from practice, timing, inhibition, and anticipatory expectations (Yarrow, Brown, and Krakauer 2009). Athletic practice strengthens cephalic expectations (Urgesi et al. 2012). Anticipatory expectations are knotted to diverse regions of the brain, including the frontal, insular, parietal, and extrastriatal cortex (Abreu et al. 2012).

Neural plasticity is revealed through diverse events, including practice in sports. Such changes in the brain underlie the improvement in physical capability and hence an improvement in sport (Ajemian et al. 2013). In other words, these changes in the neuronal patterns reflect changes in an athlete's performance (Adkins et al. 2006). The positive interaction between practice and the stabilized motor skill is demonstrated in figure 7.2.

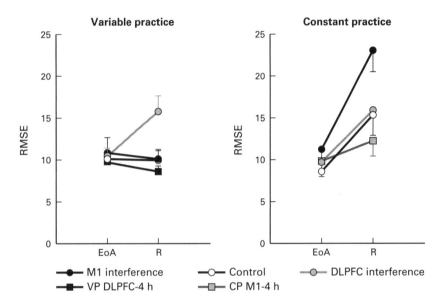

FIGURE 7.2

Practice structure and offline motor-memory stabilization.

EoA and retention error (RMSE) of participants in the control, M1 interference, DLPFC interference, and delayed interference (VP-DLPFC 4 h and CP-M14 h) groups. Left, rTMS interference to DLPFC (gray-filled circle) but not to M1 (black-filled circle), after variable practice attenuated offline motor-skill stabilization between EoA and retention compared with the control group (open circle). Performance stabilization between EoA and retention was attenuated only when rTMS was applied over DLPFC immediately after variable practice but not when applied four hours postpractice. Right, immediately after constant practice, rTMS to M1 (black-filled circle) but not to DLPFC (gray-filled circle), attenuated offline stabilization of the motor skill compared with the no-rTMS group (open circle). Performance from EoA to retention was significantly attenuated when rTMS was applied over M1 immediately after constant practice (black-filled circle) but not when applied four hours postpractice.

Source: Adapted from Kantak et al. (2010).

One study found that, when looking at groups of basketball players compared to control groups, there were gray-matter (GM) volume differences in the vermian lobule VI-VII of the cerebellum. Another study compared the structural plasticity of golfers and nongolfers, finding that golfers had larger GM volumes in premotor and parietal cortices and smaller fractional anisotropy (FA) along the internal and external capsule and the parietal operculum (Jancke 2009).

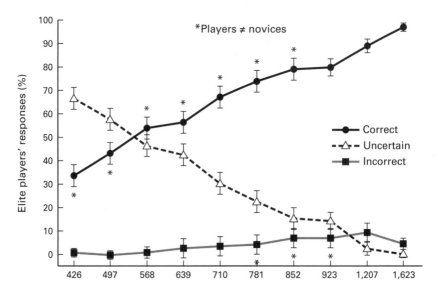

FIGURE 7.3

Skill related to subjects' predictive capability.

Percentages of uncertain, correct, and incorrect responses (mean ± SEM) made by the elite player and novice groups at the different clip durations. The percentages of uncertain responses indicate the response criterion used by each group. The percentages of correct and incorrect responses indicate the ability of the two groups to predict the fate of the basket shots. Note that the point of intersection between uncertain and correct response curves represents the clip duration at which correct responses were higher than uncertain responses. This occurred after 568 ms for elite athletes and after 781 ms for novices. Error bars indicate standard errors. Asterisks indicate significant comparisons (P of 0.05) between elite athletes and novices.

Source: Adapted from Aglioti et al. (2008).

Indeed, and not surprisingly, team-sport athletes can predict the motor response and success of fellow players, as shown in combined psycho-physical and fMRI testing (Aglioti et al. 2008). Basketball players, for instance, are better able to predict the moves of fellow basketballers than are other athletes with comparable visual/motor capability on simulated tasks. Also, the greater the capability of the basketball player, the greater his or her capacity to predict the success of others at the free-throw line.

THROWING

The arm, with its agility and flexibility, figures large in our biological and cultural evolution. Its muscularity, strength, and dexterity underlie diverse expressions and are essential in the tools that we used and developed. Consider throwing.

Accurate throwing is an ancient adaptation, part of our evolutionary legacy and perhaps linked to bipedalism (Isaac, Leakey, and Behrens-meyer 1971), although all primates tend to throw things, from rocks to sticks to their own feces. Over ten thousand years, such capabilities as accuracy in throwing found expression in hitting a target.

Human evolutionary history shows that throwing is a feature of our species and that to throw well was to survive (Roach et al. 2013). The capacity to store energy and release it with control, rapidity, and flexibility probably emerged with *Homo erectus* about two million years ago, along with greater flexibility of the torso; the infusion of energy vital for hunting and running emerged with shoulder flexibility and control over the elbow and wrist. Such flexibility and energy capability in the shoulder figured importantly in hunting. It then was filtered through play and eventually into sport.

One hypothesis is that flexibility and use of energy resulted in the power of the throwing capability (Roach et al. 2012). Cocking the arm and throwing something reflects this. Figures 7.4a, b, and c depict throwing capability and describe the flexible shoulder and its evolution. Lower humeral torsion also contributed to this evolutionary step, resulting in greater throwing capability (Roach et al. 2013). And lower humeral torsion is an important feature of athletes whose sports require throwing ability. Rotational range and speed are features of any great baseball pitcher, for instance.

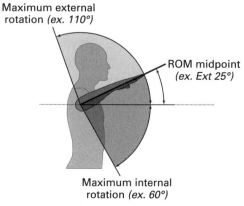

Maximum external
rotation *(ex. 110°)*

ROM midpoint
(ex. Ext 25°)

Maximum internal
rotation *(ex. 60°)*

$$\text{Torsion predicted} = 180° - \left(\frac{\overline{X}\,\text{max Ext ROM} - \overline{X}\,\text{max Int ROM}}{2} \right)$$

(ex. Torsion predicted = 155°)

FIGURE 7.4.A

Torsion is predicted from mean range-of-motion (ROM) maxima data. The difference between external and internal rotation (in this example, 25°) is equal to the ROM midpoint. The use of the 180° term allows the torsion value to be reported according to prior convention. Following the clinical definition, external rotational ROM is illustrated in light gray and the internal rotational ROM is illustrated in dark gray.

Source: Adapted from Roach et al. (2013).

FIGURE 7.4.B

Model of elastic energy storage.

(a) Arm-cocking and acceleration phases of the overhand throw. Light-gray text boxes show the relative timing of the "cocking" motions; dark-gray boxes indicate the relative occurrence of the opposing "acceleration" motions. Short boxes illustrate variation in timing of onset and cessation. (b) Humans (right) and chimpanzees (left) differ in arm abduction and elbow flexion during throwing. (c) In humans, aligning the long axis of the humerus with the major axis of the pectoralis major and flexing the elbow maximizes inertia to shoulder flexion torque and loads the elastic elements in the shoulder. However, in chimpanzee morphology there is a conflict between maximizing humeral rotation and maximizing elbow extension. Hence, chimpanzees are unable to achieve the same elastic energy storage. (d) Signatures of shoulder orientation found in the scapula (human, right; chimpanzee, left) can be used to reconstruct hominin shoulder orientation.

Source: Adapted from Roach et al. (2013).

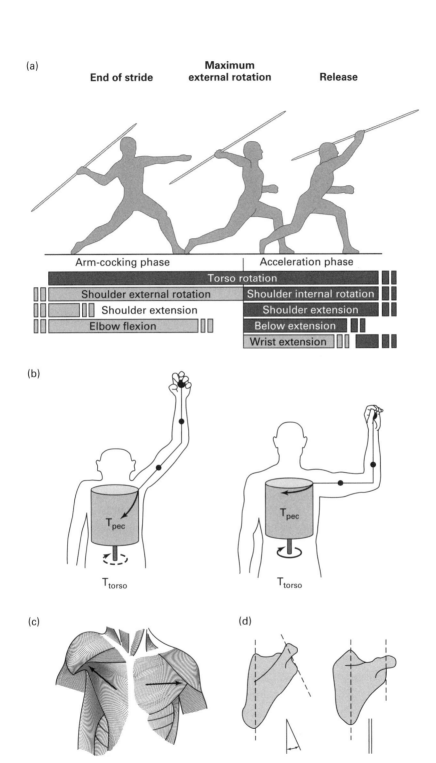

(a)

End of stride **Maximum external rotation** **Release**

Arm-cocking phase | Acceleration phase

Torso rotation

Shoulder external rotation | Shoulder internal rotation

Shoulder extension | Shoulder extension

Elbow flexion | Below extension

Wrist extension

(b)

T_{pec} T_{torso}

T_{pec} T_{torso}

(c)

(d)

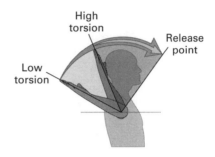

FIGURE 7.4.C

Humeral torsion and throwing performance.
Low humeral torsion shifts the shoulder rotational ROM externally.

Source: Adapted from Roach et al. (2013).

One fundamental human tool was a pointy or sharp-edged object used to carve meat and, eventually, to help in the hunt and as weaponry. Some chimpanzee groups occasionally use pointed sticks to kill small prey, so thrusting spears were likely a very early element in the human toolkit. Stone-tipped arrow use has been dated back 64,000 years; the size and shape of the stone is a piece of cultural evolution.

The javelin is a modern expression of the spear. Killing at a distance was enormously helpful in our early hunting practices, allowing us to avoid close contact with large and potentially dangerous animals (as any modern hunter will tell you, even a small deer can inflict a nasty, even bone-breaking kick). Neanderthals, unlike *Homo sapiens*, seem to have used only thrusting spears, and wounds on Neanderthal remains similar to those sustained by modern rodeo riders indicate they may have physically tackled their prey. The javelin, and eventually the atlatl (spear thrower), allowed our ancestors to avoid that kind of damage. While the sport of javelin is still a demonstration of strength and accuracy, its use is also symbolic in sports, linking human culture to biological, mental, and physical expression.

Such is the tool that is commonly referred to today as a spear. Tracking prey and then tagging with accuracy is the common currency of our species; from hunting to sports is an easy step. Of course, that step is separated in time by evolutionary events. But even early humans likely

FIGURE 7.5

Spear use. The use of spears for hunting is no longer frequently practiced; instead, spears are used in competitive sports, seen here in this photo of a competitive javelin thrower. Spear hunting, a typically social activity for early humans, remains a social activity for spectators and the athlete.

Source: Photo by Adam KR, Halifax, UK.

FIGURE 7.6

New York Yankee Mariano Rivera pitched a scoreless ninth inning against the Baltimore Orioles, July 29, 2007, at Camden Yards in Baltimore. The Yankees won 10–6.

Source: Photo by Keith Allison.

practiced their throwing abilities, and that practicing would have easily evolved into sporting competitions.

Mariano Rivera pitching: the elegance of the form and the dominance of and accuracy of the throw is a modern spectacle of a confluence of capabilities coming together in a crescendo.

WATER, SWIMMING, WATER HUNTING, AND PLAY

Vision is not everything. Consider dolphins. These mammals that returned to the water demonstrate an evolution in which cortical expansion that does not involve visual acuity is a core feature of social contact. Dolphins and whales are social animals with elaborate acoustic signaling systems.

Dolphins, known for their elaborate and long-term social bonds and complex social alliances, have a highly evolved auditory cortex and cerebellar cortex (O'Connor 2007). They emit acoustic signals that are varied, rich, and linked to diverse social cohesion. In other words, dolphins see the world through their auditory system, and this is reflected in the size and complexity of their auditory cortex.

If dolphins could communicate with us and had innate syntactical predilections about objects linked to audio, then perhaps they could do what blind children appear to be able to do. Interestingly, even children blind from birth readily understand spatial objects and concepts (Landau and Hoffman 2005). The sensory systems are pregnant with cognitive resources, emboldening the organization of action vital for the formation of social contact and social aversion.

Dolphins also display a broad array of play, acrobatic, and exploratory behaviors. The motor display and cognitive capability inherent in their behavioral play and social coordination remind us of ourselves at our best in diverse forms of social sport.

The book *From Fish to Philosopher* (1953), by Homer W. Smith, echoes the primordial role of water in our evolution and our lives. We evolved first in water, it is essential to our survival, and we have always been drawn to it. The Pre-Socratic philosopher Thales believed water was the originating principle of nature. It is not surprising that water is reflected in our sports; vast arrays of them are played in the water.

Often when we think of swimming, our mind goes to its primordial roots; we emerged from water. Indeed, the planet is mostly covered with water. It's what bathes all end organ systems and makes up all plasma fluids, in which peripheral organs are bathed and through which signals are transmitted. It is found in the cerebrospinal fluid, in which the brain rests, as though it is sitting in a pond.

Indeed, the diverse forms of swimming in sport are embedded in a larger sense of water tied to both the planet's evolution and our own. We gained limbs by crawling on land; these features were selected through terrestrial pressures that required land animals to possess appendages that enabled them to navigate nonaquatic landscapes.

It is not surprising that babies swim easily, having only just left an aquatic world. And it is not surprising that one can live longer without food than one can without water. Indeed, while we are not the only primates to swim, and certainly many animals swim and do so well, swimming is a basic human skill in most cultures.

A number of physiological features allow us to swim relatively efficiently. Unlike chimpanzees, who are terrible swimmers and sink like stones, we humans have a layer of body fat that makes us more buoyant in water than most primates. We also have the mental and physiological capacity to hold our breath under water, sometimes for extended periods. Such capabilities reflect both biological and cultural evolution. Our upright stance also makes us more streamlined in cutting through the water; perhaps standing erect put us in a position to exploit more diverse resources in water as well as on land.

There are four competitive swimming strokes: freestyle (competitors are allowed to choose any stroke, but typically the front crawl is chosen), backstroke, breaststroke, and butterfly. The front crawl involves alternating arm strokes over the surface of the water with a flutter kick while facing the bottom of the pool; the backstroke involves the same arm and leg patters while the swimmer's back is to the bottom of the pool. Breaststroke consists of simultaneous movements of the arms in the horizontal plane: the swimmer's hands push out from the breasts in a heart shape while the legs kick in a motion similar to that of a frog. Finally, butterfly consists of simultaneous movement of the arms in circular motion above the surface of the water with a dolphin kick; it is often perceived as the most beautiful stroke. The various forms of competitive swimming include: fifty-meter freestyle, short-distance freestyle, middle- and long-distance freestyle, medley, relay, diving, synchronized swimming, and open water.

Preadaptive capabilities resulted from the biological breakthroughs that enhanced our cultural evolution. We often drive ourselves to exhaustion, an act that elite swimmers sometimes call being "broken down." This tolerance for exhaustion and the ability to push through it likely originates in the need to hunt or avoid being hunted, both in and out of the water.

Whether you were a hominin at home in the trees or one that relied on navigating the ground alone and traversing vast landscapes, access to water and what it provides—both as a necessary resource and another landscape to dominate—is a constant that is tied to endless cultural evolution.

Our capacity to explore and pursue resources is tied to the boats we build and the places we explore. Hunting in water is as commonplace as hunting on land. We probably started by foraging in the shallows, picking up crabs, oysters, and clams. Shell middens, heaps of discarded shells from edible sea life, are found in notable quantities throughout the world at archaeological sites dating from the Late Mesolithic period (four thousand to ten thousand years ago). The oldest shell middens, in South Africa, are about 140,000 years old (Hirst 2014).

Perhaps an early human, out beyond her depth, grabbed on to a floating tree trunk. It wouldn't have taken much for our tool-using species to have figured out a better mode of water transport and to have adapted tools for catching fish. Cave art in southern France shows harpooned seals, and harpoon points suitable for spearing fish have been uncovered in the Katanda region of Zaire, dating from eighty thousand years ago.

Rafts are hypothesized to be some of the earliest forms of nautical transport. While no direct archaeological evidence exists, indirect evidence such as stone tools and genetic migration patterns suggest that such rafts may date back to *Homo erectus* and the peopling of Australasia. There is some indirect evidence of the use of hide boats during the Bronze Age in the archaeological record: cave paintings in Russia and Scandinavia are thought to depict hide boats, but they are difficult to date and interpret. Log boats have been found and dated with much more success. The oldest surviving examples, which date to around 7000 BC, were found in Pesse, Netherlands, and Noyen-Sur-Seine, France. These boats could hold up to eighteen passengers, depending on the size of the tree used to carve the boat. Finally, the oldest plank boats excavated to date were built between 1900 BC and 400 BC. They were built with wooden planks, often fastened together with flexible yew withies.

Exploration also encouraged us to exploit our capabilities on the water. When humans colonized Australia sixty thousand years ago, sea levels were lower than they are now, but it still would have been nearly a day's voyage from what is now Java to the north coast of Australia. Land would not have been visible during the middle of the voyage; those early people must have been astute observers of cloud formations and bird movement to know that the continent was even there. And they must have traveled on boats, rafts, or dugouts of some kind.

It is easy to see how hunting skills and diverse sports are linked, as shown in the evolution of the use of spear from hunting to javelin throwing. (Walls 2012). Today, there are a number of Olympic rowing events, with both men and women competing in single sculls, double sculls, quadruple sculls, coxless pair, and eight events, with men adding the coxless four, as well. Kayak games and hunting are essentially linked in various cultures across the globe. Skill, constant practice, and accuracy are tied to cephalic capabilities and cultural context. The Inuit, whose modern origins are in the Bering Strait, eventually migrating from Siberia to Greenland, reveal the continuity of hunting and sport and the development of embodied sensibility and excellence in the water (Walls 2012).

CONCLUSION

This dense, interdependent cognitive network we call humanity is really a fairly recent phenomenon. We—modern *Homo sapiens*—emerged only about one hundred thousand years ago. Agricultural society and the domestication of food animals such as sheep and goats, the move away from a hunter-gatherer state, developed in the range of ten thousand years ago. The creation of written language emerged something in the order of five thousand years ago: this is so amazingly close in time to where we are yet seemingly so distant from the complexities of contemporary life (Tattersall 1993). Formal sport and its rules also seem to have emerged over this

TABLE 7.1 Early Old World localities with possible evidence for aquatic resource use

LOCALITY/SITE	AQUATIC FAUNA AND ASSOCIATIONS	AGE (YR)
Archaic *Homo sapiens*		
Hoxne, England	Remains of fish, otter, beaver, and waterfowl associated with Acheulian deposits; distributions similar to artifacts, suggesting a cultural origin	~350–300K
Duinefontein 2, South Africa	Sea bird (penguin, cormorant) remains in Late Acheulian site dominated by land mammal bones	~400–200K
Lazaret, France	Marine shellfish in Late Acheulian context	~186–127K
Kebibat, Rabat, Morocco	Aterian shell midden on Atlantic coast, associated with Neanderthal remains	150 ± 50K
Anatomically modern humans (*Homo sapiens sapiens*)		
Herolds Bay Cave, South Africa	Early Mesolithic shell midden with mussels (*Perna perna*), other shellfish, and otter remains associated with hearths	~120–80K
Katanda 9 and 16, Semliki River, Zaire	Thousands of fish bones associated with MSA barbed bone harpoon points in riverine setting	~90–75K
Willandra Lakes, Australia	Abundant shellfish and fish remains from numerous lakeside camps, associated with terrestrial fauna and mixed economy	~50–15K
Kilu Rockshelter, Solomon Islands, Melanesia	Shell midden with fish bones and other fauna; colonization of island required several substantial voyages by maritime peoples	29–20K

Note: M = million years; K = thousand years.

Source: Adapted from Erlandson (2001).

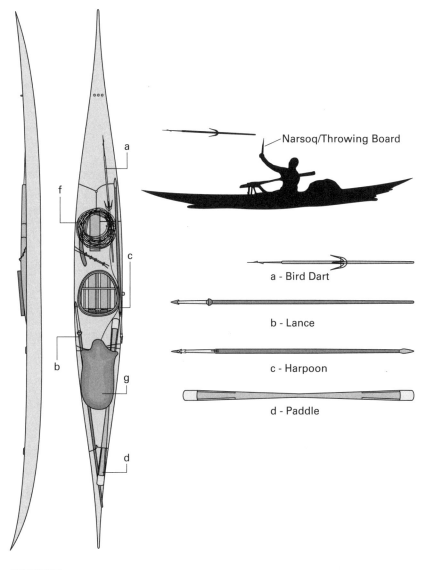

FIGURE 7.7

Greenlandic kayak from the 1800s and the array of hunting weapons that were carried on the deck during normal hunting expeditions. This kayak is now located at the Danish National Museum in Copenhagen.

Source: Adapted from Walls (2012).

most recent five-thousand-year period (Guttmann 2004). Our capabilities are written across all things cultural, and sport stands out easily.

Elite athletes are a special brand of human being. Athletic capability is a thing of great beauty, efficiency, and excellence, and these superlatives all have a common denominator: brain expression. For instance, the amplitude of alpha rhythms in elite athletes, when compared with either amateur athletes or nonathletes, is more integrated. Therefore, focus is heightened, rest is potentiated, and information processing may be enhanced (Babiloni et al. 2010).

The enhanced processing of information pervades all sports as a critical feature of diverse forms of adaptation. For example, such anticipatory controls pervade the motor systems, as evidenced by the simulation studies that show the degree to which elite cricket players anticipate the bowler's intentions (Muller, Abernethy, and Farrow 2006). Anticipating the intentions of others, sometimes called holding a "theory of mind" (Premack 1990), is a fundamental feature of our species.

Sport is a rich panoply of biological continuity with cultural expression; the one blends endlessly into the other. Evolution is not a closed system, and sport is an easy marker of such blending. There is no overt determination of most behaviors, but there is the endless creativity, the construction of our tools—in this case our sports tools, the core capabilities that render this possible, and our creative capacity to get better and better at our activities.

Of course, cultural expression exists within biological constraints; in an earlier philosophical piece I called this "possibilities within constraints." There are real facts of the matter about space and time, oxygen consumption, height and weight, and the rest. It is just that our cephalic capabilities expand bodily expression within a context of cultural evolution. We celebrate the expression while we determine the extent of our plasticity—in this case, our possibilities. Our biological capability is expressed through sport, and sport expands our mental and physical capabilities.

8

FAIRNESS AND SPORTS

Moral sensibility runs through sports. After all, what defines our evolution in general is a conception of morality, whether thought of in the context of the state, tribe, team, or individual. Although brains vary with species and taxa, linking a conception of evolution with brain function requires a conception of the nervous system in its core outline, variation, and common themes (Cajal 1906, Swanson 2000).

The neurologist J. Hughlings Jackson (1884) formulated a conception of the nervous system that considers the brain within the context of evolution and in terms of levels of function. We now think in terms of distributed systems across the neural axis, in which information molecules are prominently expressed and play diverse roles in the organization of action.

Development of the neocortex was a crowning achievement for our species, greatly increasing our range of action, yet many of the effects of practice and sport occur outside the cortex. The received view of the cortex is that it restrains behavioral and physiological systems: teenagers with raging hormones and cortices that are not fully developed are like decorticated animals in this view, lacking inhibition. But that is a one-sided view. The neocortex is also linked to restraining amygdala function and to the enculturation process, which is so vital in sport, for instance, in determining when to exert and when not to. The restraint of behavior, the taming of wants, is vital across human life and noticeably in sport.

Cortical expansion is linked to deception in addition to cooperation. The practice of deception is significant in human cultures, and a number of comparative studies have highlighted it as a feature of primate evolution.

In this chapter, which focuses on our cultural evolution through the participation in sports, we will begin with a discussion of drug abuse in sport before moving on to consider fairness in sport and the expansion of our sense of rights and human participation.

Deception: At a young age we learn about the consequences of not doing what is right and about the importance of truth telling in sport, and one aspect of that is learning the consequences of deceit. The difference between truthfulness and deceit can be murky. Nations and teams want to win, and winning, like eating, is a driving force. Deceiving or lying may serve that end. The ability to mislead is a primary feature in ontogeny. Skill at predicting events and the reactions of others easily leads to the manipulation of those skills to get what one wants at any cost, including that of cheating (Greene 2014a, 2014b).

A number of brain regions are linked to lying and cheating. I don't believe they are there simply for the purpose of expressing deceit (although others do). To me, they seem to underlie diverse forms of social and problem-solving capabilities. After all, deceit is one route among many to achieving goals. Misleading others appears to be an ancient adaptation. Sarah, a chimpanzee who worked with the Premack group a number of years ago, would mislead the trainer that she was less friendly with or attempt secretly to aid the trainer she liked in object-search tests (Premack and Woodruff 1978). Indeed, the greater an animal's social capability, the more it finds deception useful in a social context for the exploitation of resources.

In our species, sociopaths use an extreme version of lying and cheating, which is often linked to what psychiatrists call "cold sensibility." In neural terms, this is a disconnection between visceral feedback and decision making; practically speaking, sociopaths mislead others with no feelings of compunction, even when their behavior directly harms others; there does seem to be a continuum of behavior regarding what individuals regard as acceptable levels of deceit.

Children have to learn to be truthful: they need reinforcement for truthful behavior in order to develop what we think of as a conscience.

Some people pick this up more readily than others, and different cultures reinforce it to a greater or lesser degree. But truthfulness is often a "faint motive" even under conditions where truth telling is supposed to be a primary feature (for example, the practice of science). We have cultural mores to promote and then check truthful behavior, and those apply in sports. As President Reagan said about nuclear disarmament, "Trust, but verify." In sport and in life, we look to verify. The marker of our cultural evolution is the speed and accuracy of such verification events.

Prefrontal and frontal cortex inhibition of diverse regions of the brain, including the amygdala, underlie decision making about cheating and lying (Bechara 2005, Greene 2014b). The general structure, which is embedded in the crevices of the cortical/amygdala region, relies on structural connectivity and the diverse information molecules that underlie those connections. One result is behavioral expression: telling the truth or not, lying or not, playing fair or not.

Coming back on a plane from a committee meeting, a weightlifter and I got into a conversation. He admitted that, going back to the late 1960s, they all took anabolic steroids to get strong. Use of anabolic and catabolic steroids is still a constant concern for those supervising bodybuilders and their competitions.

And of course, there is the cultural cops-and-robbers evolution: as the cops figure out how to detect illicit drug use, the robbers get better at using old drugs and discovering new, undetectable ones. From simple tricks such as substituting another person's urine for a drug test, users have graduated to diverse forms of masking, such as timing when to take the drug. That requires knowing a lot about how not to get caught. It demonstrates how effective we as a species are at deceiving. And when the deception is institutionalized and has a tradition (as it was with weightlifting and as we now see in the cycling community), it becomes extremely difficult to root out.

The good news is that deception is only one human cephalic capability among others. Fairness also matters to us, and sport has a long tradition of emphasizing fairness and, both literally and figuratively, an even playing field. It may be tenuous, but it is inherent; we come prepared to measure and keep track of fairness. It's not an accident, for example, that sports led the way in breaking the race barrier in the United States.

STEROIDS AND PEPTIDES: DIVERSE INFORMATION MOLECULES

Endocrines (for example, steroids and peptides) are what I call information molecules. A partial list of major hormones secreted by the endocrine glands are growth hormone, prolactin, adrenocorticotropin, thyroid-stimulating hormone, oxytocin, endorphins, thyroxine, glucocorticoids, androgens, epinephrine, angiotensin, vitamin D, insulin, estrogen, testosterone, cytokines, and leptin.

The major neuropeptides include atrial natriuretic factor, substance P, interleukin, β-endorphin, urocortin, vasotocin, calcitonin, angiotensin, gonadotropin-releasing hormone, dynorphin, corticotrophin-releasing hormone, vasoactive intestinal polypeptide, galanin, vasopressin, luteinizing-hormone-releasing hormone, somatostatin, neurotensin, prolactin, bombesin, neuropeptide Y, somatostatin, enkephalin, and thyrotropin-releasing hormone.

Some of these chemical signaling systems in the brain are the neurotransmitters we previously discussed, including dopamine, norepinephrine, and serotonin. These are broad-based chemical signals that regulate broad-based behavioral and physiological responses. But other information molecules have more specific roles. They developed in the context of evolution, and they underlie sport activity.

One must keep in mind the blood-brain barrier. Information molecules such as peptides are kept out of the brain when they are synthesized in other parts of the body; when they are produced in the brain, we call them neuropeptides. Steroids are mostly not produced in the brain, but, being lipid permeable, they can get directly into the brain.

All steroids are derived from cholesterol. They are mostly produced outside the central nervous system, though low levels of progesterone are produced in the brain (Schumacher and Robert 2002). Importantly, steroids cross the blood-brain barrier (unlike peptides). Common ancestral genes (Evans 1988) underlie diverse steroid hormones; one such common relationship is to an ancestor estrogen receptor (Thornton 2001). Progestereone, cortisol, testosterone, and thyroxine have a common molecular ancestor, but they also have specific receptors in which there is differential competition and specificity.

Anabolic steroids are mostly linked to sport and aggression, muscle expansion, libido, and competitiveness across species; they are phylogenetically ancient, as are most information molecules. Analogues may be found even in invertebrates. While doping scandals have given anabolics a bad name, they are essential in the organization of action in general and sport in particular.

The Antiquity of Information Molecules

Life on this planet began an enormously long time ago. The formation of the earth dates to around 4,500 million years ago. Small bacteria emerged some 3,000 million years ago, single-celled algae some 2,000 million years ago, and jellyfish appeared 550 million years ago. Plants, dinosaurs, and mammals evolved over the last 500 million years.

This change seems slow, compared with the mere 100,000 years it has been since the rise of modern *Homo sapiens*. But it took millions and millions of years to get to how things are today. The period of time encompassing the evolution of our hominin ancestors is so short yet so profound in human change, but change is the underlying theme in evolution, along with adaptation. Variability of outcome promotes long-term longevity.

Species come and go, but one common thread is the persistence of diverse information molecules such as peptide hormones. Endorphins, linked to pain perception and euphoria in athletes, are expressed in diverse living things beyond mammals or vertebrates more generally. These are ancient molecules.

In mammals, these information molecules became oxytocin and vasopressin. Many of their functions are originally knotted to fluid balance (Fitzsimons 1998), but they diversified in function (Pickford and Strecker 1977, Strand 1999, Gimpl and Fahrenholz 2001). Many of these information molecules are ancient in function; consider gonadotropin-releasing hormone (Sherwood and Parker 1990, Bentley 1982), which dates back to jawless fish. Antiquity and diversification are common features across the basic information molecules in the brain (Strand 1999).

Indeed, the peptides CRH (cortisol regulation) and TRH (for example, thyroxine regulation), which were named after being found to be part of the HPA or TSH (thyroid-stimulating hormone) axis, respectively, underlie the physiology of change. One dramatic change is metamorphosis in animals that are both land and water dwelling (for example, the frog). These same information molecules underlie pregnancy in humans (Power and Schulkin 2009). Sport is rich in adapting to change, that is, in anticipating and predicting events; such capability is built into the information molecules across the range of bodily tissue.

ANABOLIC STEROIDS AND THEIR ABUSE

Anabolic steroid abuse is common, and some of it is state sponsored. The East Germans in the 1970s and 1980s were notorious for this, but other countries and also some respected athletics clubs made it almost into an art—an art of denial and not getting caught—before widespread testing began to reduce somewhat its excessive and overt use in professional and higher-level amateur sports.

Importantly, the brain is estrogenic: an aromatization process converts testosterone into estradiol or estrogen-like compounds in the brain (Goy and McEwen 1980). This has a long-term structural impact on the brain and on behavior. However, there are receptors for estrogen beyond those in the gonads or for testosterone beyond those in the testes; sexual dimorphism is apparent at diverse points in the brain, including the medial nucleus of the amygdala, medial bed nucleus of the stria terminalis, and medial preoptic region. Levels of testosterone or its metabolites are a clue in detecting abuse of this steroid. Indeed, in this age of increased acceptance, who counts as a female also has a curious history in relation to participation in Olympic sports (Wood and Stanton 2012).

What we want is transparency and fairness; when we discover that a male is disguised as a female in order to gain a competitive advantage, we are offended, and we should be. Though males who compete as females are playing unfairly, we are witnessing an expanding body politic in our culture: a broader range of individual and group variation is being included in our social context.

In 1989, about 10 percent of American male adolescents abused anabolic steroids. Over the next few years, the rate increased to around 50 percent of males in organized sports in general and 80 percent of weightlifters.

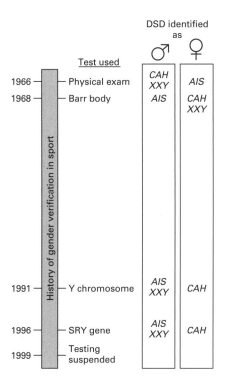

FIGURE 8.1

Timeline of gender verification testing, showing the method used and the resulting gender for individuals with selected disorders of sexual differentiation. AIS = androgen insensitivity syndrome; CAH = congenital adrenal hyperplasia; SRY = sex-determining region on the Y chromosome; XXY = Klinefelter's syndrome.

Source: Adapted from Wood and Stanton (2012).

Perhaps the most abused illegal enhancers are testosterone-related molecules (Wood and Stanton 2012). Anabolic steroids' ability to increase muscle mass is well known: compare the arm circumferences of early baseball greats such as Hank Aaron and Lou Gehrig with those of Barry Bonds or Roger Clemens.

A lesser-known result of steroid abuse is physiological breakdown, including of liver and cardiovascular function, in both men and women. Steroids are also tied to psychological dependence and addiction

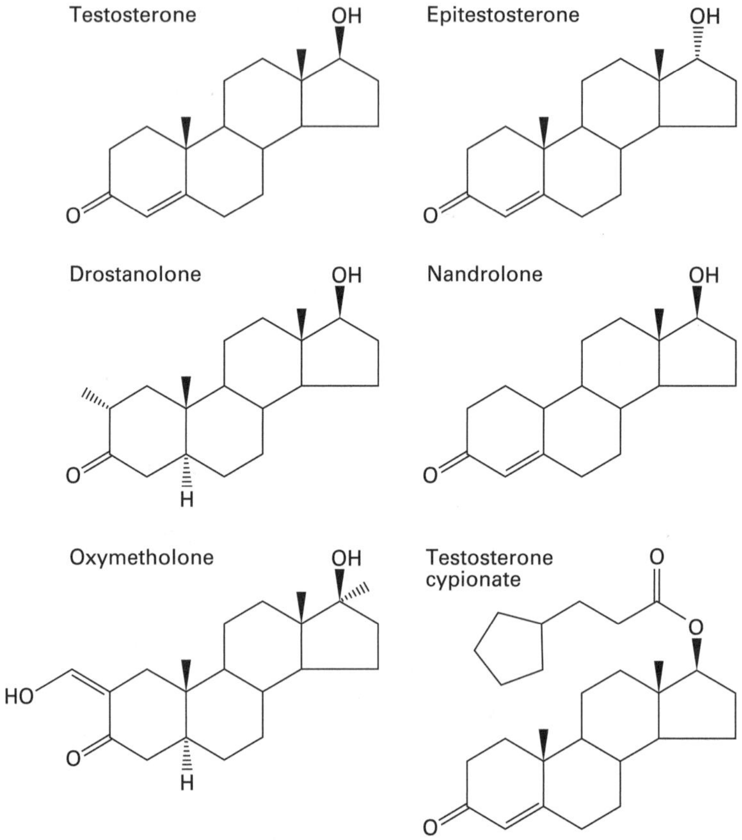

FIGURE 8.2.A

Molecular structure of testosterone, epitestosterone, and some popular anabolic-androgenic steroids.

Source: Adapted from Wood and Stanton (2012).

(Quaglio et al. 2009), among other problematic effects. While anabolic steroids do have the commonly known effects of increasing body mass, reducing deep intramuscular fat, increasing the volume of muscle fibers, and increasing the force and power of muscular contraction, their continuous use also leads to quite a few complications. The genitourinary and reproductive systems, the metabolic system, the digestive system, the skin,

Percentage of positive tests

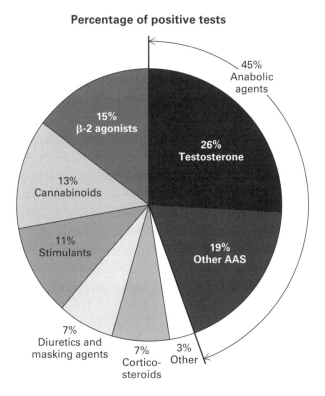

FIGURE 8.2.B

Percent of adverse analytic findings reported by the World Anti-Doping Agency in 2006.

AAS = anabolic-androgenic steroids.

Source: Adapted from Wood and Stanton (2012).

and the skeletomuscular system are all negatively affected. Steroid users also suffer neurological complications like cephalea, drowsiness, and a loss of concentration as well as psychiatric complications like aggressiveness, irritability, depression, manic behavior, and paranoid psychoses.

Anabolics cause a wide range of psychological effects, facilitating a range of emotions from euphoria to despair and anxiety and altering aggression levels, libido, fear, and appetite. In experiments with mice, anabolic steroids potentiate fear-related behavior; mice show a higher startle response in anticipation of shock as a function of treatment with anabolic steroids.

One information molecule, CRH, is tied to responses to danger (Schulkin 1999). Anabolic steroids appear to increase CRH gene expression in regions of the brain such as the central nucleus of the amygdala. The bed nucleus of the stria terminalis, part of the head ganglia of the HPA axis, is active in regulation of internal milieu (Herman et al. 2003). Many information molecules are produced in this region, including CRH and endorphins, which play a role in various behaviors. Steroids, including the anabolics, induce the production of neuropeptides such as CRH, which results in anxiety (Costine et al. 2010). These information molecules are tied to both adaptation and pathology.

The increased aggression produced by anabolic steroid supplements may initially seem to provide another advantage for athletes. But aggression is not a unitary concept or state of the brain: there are many kinds of progesterone- or oxytocin-induced aggression. Maternal aggression is one kind, territorial aggression another. Not all forms of aggression are advantageous in sport, and excessive aggression can counterbalance self-control, patience, and precision.

The various forms of aggression share some common elements. One is steroid facilitation of neuropeptide gene expression, for which testosterone- or estrogen-induced vasopressin in the medial amygdala is one model. Here, the aggression is for territorial defense (Neumann and Landgraf 2012). Testosterones and their metabolites, which have long been known to enhance aggression and impulsivity, are linked to steroid dependence, a form of addictive behavior (Wood and Stanton 2012; Herbert 2015).

It is common knowledge that testosterone enhances aggression (Wood and Stanton 2012), but independent of testosterone-induced changes in morphological features (muscle mass, facial features), enhanced physical aggression in general is expressed early on in differences between boys and girls in rough-and-tumble play. But, of course, one fundamental function of testosterone is reproductive capabilities (Herbert 1993, 2015). Testosterone, like estrogen, is widely distributed across the brain and is phylogenetically ancient, and it is found in both vertebrates and invertebrates.

Anabolic steroids are tied to epigenetic events, that is, to the enhancement and decrement of diverse neuropeptides and neurotransmitters in the brain and in the body in general. Diverse genes are turned on and off by DNA methylation.

Testosterone levels may rise naturally after winning a sporting competition and decline with losing (Herbert 1993, 2015). Diverse forms of risk taking have also been linked to testosterone levels. Though this is not axiomatic, the anabolic steroid hormones do set the conditions that, under context and culture, need and temperament, lead to the expression of aggression. Indeed, steroid induction of vasopressin in regions of the brain such as the medial amygdala can result in diverse forms of competitive behavior.

Anabolic steroids, in addition to a number of endorphins, influence central dopamine expression, which might link their use to addiction (Herbert 1993). In other words, anabolic steroids, by the induction of dopamine or central opioids, render the athlete or the bodybuilder susceptible to addictive forms of behavior.

Athletes are by now fairly well educated about the long-term problems associated with steroid abuse, and coaches know what signs of abuse to look for. But sports figures try to exploit a panoply of other drugs, from Hepatocyte Growth Factor [HGF] to beta blockers, from diuretics to blood doping (drawing blood and then reinjecting it before a game). Dope testing, education, and awareness have changed how athletes, officials, and the public approach the topic, but it has not halted drug abuse in sports; truth is often a faint motive, and it competes with other motivations, namely winning. Many professional and amateur sports organizations have banned the use of various supplements. Figure 8.1 represents a subset of the supplements that the International Olympic Committee (IOC) and the National Collegiate Athletic Association (NCAA) have allowed and prohibited for their participating athletes.

Now, with so much doping, what about ethics?

EUGENICS

Eugenics literally means "good birth": Under the influence of eugenics, depictions of evolutionary racism were commonplace. Georges Cuvier

Creatine

This compound, which occurs naturally in the body, is sold as an over-the-counter supplement primarily to enhance recovery after a workout and increase muscle mass and strength. Creatine is popular with football and hockey players, gymnasts, and wrestlers. Side effects include weight gain, nausea, muscle cramps, and kidney damage.

Anabolic Steroids

Anabolic steroids (also referred to as anabolic-androgen steroids) are synthetic versions of the hormone testosterone, and they are used to build muscle mass and increase strength. They're popular with football players and weightlifters. Side effects include heart and liver damage, halted bone growth, and a permanently short stature.

Anabolic Agents (Including Testosterone)

The primary medical use of these compounds is to treat delayed puberty, some types of impotence, and wasting of the body caused by HIV infection or other muscle-atrophying diseases. Some physiological and psychological side effects of anabolic steroid abuse can affect any user; others are gender specific.

Steroid Precursors

Steroid precursors, such as androstenedione ("andro") and dehydro-epiandrosterone (DHEA), are substances that the body converts into anabolic steroids. Androstenedione is produced by the adrenal glands, ovaries, and testes and is normally converted to testosterone and estradiol in both men and women. Steroid precursors are used to increase muscle mass. Most are illegal without a prescription. DHEA, however, is still available in over-the-counter preparations. Though manufacturers and bodybuilding magazines tout andro's ability to allow athletes to train harder and recover more quickly, its use as a performance-enhancing drug is illegal in the United States.

Scientific studies have also shown that supplemental androstenedione doesn't increase testosterone or strengthen muscles. Most of the andro is in fact rapidly converted to estrogen. Side effects are similar to those for steroids.

Peptide Hormones, Growth Factors, and Related Substances

Medical uses of these compounds include treatment of cancer and aiding those born prematurely. The presence of an abnormal concentration of a hormone, its metabolites, relevant ratios, or markers in an athlete indicates a prohibited substance unless the person can demonstrate that the concentration is caused by a physiological or pathological condition. Examples of these substances include human growth hormone (hGH), erythropoietin (EPO), insulin, platelet-rich plasma (PRP), human chorionic gonadotrophin (HCG), and adrenocorticotrophin (ACTH). The intramuscular injection of platelet-derived preparations (such as PRP and "blood spinning") requires a Therapeutic Use Exemption. Platelet-derived preparations administered through other routes (such as local injection into a joint, tendon, or ligament) require only a USADA website Declaration of Use.

Beta-2 Agonists

The primary medical use of these compounds is to treat respiratory ailments, such as asthma. Some studies have shown beta-2 agonists have performance-enhancing effects when consistently high levels are present in the blood.

Tetrahydrogestrinone ("THG" or "The Clear")

This is the steroid purportedly used by many high-profile athletes, including the track star Marion Jones and the baseball player Barry Bonds.

Erythropoietin

This type of hormone is used to treat anemia in people with severe kidney disease. It increases production of red blood cells and hemoglobin, resulting in improved movement of oxygen to the muscles. Epoetin, a synthetic form of erythropoietin, is commonly used by endurance athletes.

Human Growth Hormone (hGH)

Also known as gonadotropin, this hormone has an anabolic effect. Athletes take it to improve muscle mass and performance. However, it hasn't been shown conclusively to improve either strength or endurance. It is available only by prescription and is administered by injection.

Source: Adapted from Kersey et al. (2012)

TABLE 8.1

SUPPLEMENTS (AND THEIR POSTULATED EFFECTS)	EFFECTS	IOC AND NCAA RULING
Caffeine (increase energy, decrease fatigue) Carbohydrate-electrolyte beverages (increase lean muscle mass) Creatine (improve muscle energy and strength) Sodium bicarbonate (increase buffering capacity)	Performance enhancing, with minimal adverse effects	Allowed
Amino acids (increase levels of growth hormone) Beta-hydroxy-beta methylbutyrate (decrease protein breakdown, increase synthesis) Chromium (improve muscle energy and strength) Iron (increase energy and general performance)	Ineffective or lack of evidence of performance-enhancing effects	Allowed
Anabolic steroids (levels of growth hormone) Blood transfusion and erythropoietin (increase endurance, oxygen delivery)	Performance enhancing, with dangerous adverse effects	Prohibited
Androstenedione and dehydroepiandrosterone (increase testosterone, lean muscle mass) Ephedrine and pseudoephedrine (increase energy and decrease fatigue) Human growth hormone (increase muscle protein synthesis, strength)	Ineffective or lack of evidence of performance-enhancing effects, with dangerous adverse effects	Prohibited

Source: NCAA, IOC

(1817), one of the first great taxonomists, had already described the Europeans as at the top of the evolutionary chart, with "Mongolians" and "Ethiopians" (his terms for Asian and African peoples, whom he believed to have smaller skulls and more monkeylike features) on the bottom. This influential view pervaded common folklore, eventually evolving into repugnant racist depictions.

The idea of eugenics did not begin with Galton (1883), even though he coined the term and was the first to apply it extensively to human development. Cattle ranchers, for instance, have bred for shape, form, and excellence since human beings began to cultivate crops and animals. The study of shape and form through cultural practices is a core cephalic adaptation.

Eugenics, the idea of maximizing a more perfect state and minimizing a less perfect one, is an ancient and essential part of our cultural history. It even corresponds to a bright spot in the sense of taking hold of our destiny. The humanistic traditions of the Enlightenment were animated by a conception of what we want to be and were linked to an experimental spirit by the Renaissance sensibility of experiment, exploration, testing, and comprehension.

Galton, however—mired in the end in controversy about the validity of his expertise and his experimental findings—turned eugenics toward the worst in us: the Nazi sensibility of the superman, the perfect being, surging with raw power and unmitigated by softness and social sensibility. This was probably a bastardization of what Galton had in mind. What Galton worried about was how to preserve and protect the very best about the British Empire.

One disgraceful instance in the United States was the exhibition at the Bronx Zoo at the beginning of the twentieth century of an African man in a cage (Newkirk 2009). This was the same zoo that helped save the American bison from extinction. This shows how generic racist evolutionary biology had become, its metaphors steeped in lethal injustice.

Of course, much of human expression, and more generally animal expression, is easily abused, and it has been. One feature of the domestication process was the size of end organ systems such as the adrenal gland (Richter 1949); experience in an enriched environment could either increase or decrease neural hybridization in certain regions of the brain. Sports that involve animals such as the horse and the dog are understood in this way.

Sport and its evolution should be one arena where these pernicious differences are diluted and where biological excellence is respected and understood. This book is about sport and its essential role in our lives within a sense of the continuity with regard to biology and culture. What should emerge from the study of sport is how culture allows for the expansion of our capabilities and for a cultural evolution that does not bury differences among us but engages them.

We are constantly shaping the contours of the body, trying to expand our capabilities. Understood in light of human performance, expression enhancement is a good. But understood and practiced in light of the denigration of human dignity, it is obviously not good. Though for the most part this is not very complicated, how far do we go? How we understand ourselves in this context of engineering, biological or otherwise, is a feature of us; we are engineers. We do it all the time in sports (and, of course, in many other milieus).

The modern Olympics began at the end of the nineteenth century. The Olympics' promotion of healthy competition between nations provided a way to show off the development of the nation-states through sport and excellence in athletics. But all human inventions are vulnerable to ill expression. Sport for the Nazis adapted glorified Roman symbols for their national emblems and cultural identity: the sacred Volk. The 1936 Olympics was to be a showcase of German excellence in sport, illustrating the superiority of the master race over lesser humans. That glorification of human perfection, dominance, and superiority denigrated the shared sense of human competition.

The 1936 Olympics was memorialized by Leni Riefenstahl in *Olympia*. Another film by Riefenstahl is *Triumph des Willens* (*Triumph of the Will*), covering the 1934 Nazi Party rally in Nuremberg. The will in question was not supposed to be about the trampling of others but the perfection of a self, indeed, about an independent self aimed at self-expression. Thus, the language was about the triumph of the will; the reference was (mistakenly) to Nietzsche.

This was understood in the context of a variant of Darwinian theory known as Social Darwinism popular at the end of the nineteenth century and in the twentieth. It stressed promulgating the fit and eliminating or reducing the less fit. Nazi promotion of "social hygiene" was interlaced with an obsession with fitness, competition, and, in some cases, brutal sport.

The Nazis went about their social program of eliminating the less fit massively and with an extraordinarily perverse attention to detail and efficiency, generating and perfecting excellence as they also distorted it. They are likely the worst possible example of a normative human drive. Perfection or aiming for excellence is the normative goal, but progress in the larger sense is about full participation. That is the larger cultural context of sport and the evolution of our species from biological capabilities to cultural opportunities.

Wanting to achieve, strive, and live are core animal predilections— perhaps "an animal faith": wanting to survive, make contact, persevere, compete, and cooperate in a social context of a team. But perfection demythologized is just being able and willing to try hard under diverse conditions. Some of us are better at trying hard than others, and some of us are really talented at diverse sports. But just about all of us can participate. Participation broadens our humanity.

In some sense, sports represents the evolution of progress and excellence in culture. The progress of a basketball team through the ranks of March Madness, for example, involves the elimination of the less-fit teams. As in evolution, the teams less adapted to the conditions of the contest do not survive.

But this world is also one of adaptation, fitting into niches and finding changing circumstances in which to participate. Cultural evolution, in at least a democratic context (Dewey 1910, 1908), is the widening of the arena of participation, not the narrowing of it. At the heart of the origins of democracy only the few participated (for example, male citizen landowners), but more and more have been included in its action as time goes by. Sport is—or should be—a great example of the widening of participation.

We include the less fortunate, using sport to embolden the sense of being in the world. What capabilities can be enhanced as features of dignity and meaning are strengthened. We are, after all, social and self-preserving; culture changes everything somewhat and some things a lot.

One thing is the extension of opportunity to the less fortunate under certain cultural contexts. It may not be within our narrow interest; it may cost. The benefits are the ideals that we can embody. And one is the sense of what it means to be more—not less—human; sport again rears its head. We can gear competition in a way that we can include others where possible. Thus, we develop sporting structures that allow for boxers, say, to fight within categories in which they can meaningfully compete and show excellence. We don't require them all to climb into the same ring; that would just not be sporting.

Our cultural evolution is a view of human's frail progress. Part of it is embodied in our sense of sports and human fairness: winning without cheating, maximizing aggression and motivation without excessive hurting, and using effective tools to protect us while we compete (finding ways to decrease concussion while preserving the game of football, for instance). This does not make sports easy to do, nor does it open all sports to everyone. There are always risks in playing, but we ought always to look for ways to decrease the propensity for injury. This will never be perfect: many sports are inherently tied to possible—and even probable—injury. But we use our technological capabilities to decrease the risk.

FAIRNESS

Engaging in play that turns into sport is one way we learn about fairness: following the rules, playing the game fairly. Mammalian play is at the root of the socialization process so essential for getting a foothold in the world, and fairness is a feature of our cultural evolution, a feature that coexists and evolved with sports participation.

In sport, rules have to be fair in the sense that they are good for all. What Rawls (1971) called a "veil of ignorance" is also part of fairness in sports: that is, we make a choice because it is fair even if we don't know whether we receive a benefit of that fairness. Of course, moral sentiments are also a piece of our biology. We are group affiliated, so where we belong and how we belong there play a role in whether we help, calm, appease, affiliate, become aggressive, or withdraw. These are parts of our cephalic equipment.

Moral sentiments—sympathy/empathy, fairness, discipline, loyalty, etc.—are social in orientation. Charles Darwin, like Adam Smith (1759) before him, viewed morality as setting the conditions for social conduct. Moral sentiments play a role in approach or avoidance, helping or hindering, and the many variants of these core moral sentiments pervade human experience.

Visceral expressions are tied to moral digressions, one prominent feature of which is disgust (Rozin 1976, Haidt 2007, Johnson 2014). Moral disgust, a pervasive expression in our lexicon, may be tied to diverse transgressions in our perception of others. The visceral nervous system runs through the brain; modern anatomy has uncovered the direct connectivity of forebrain and brainstem sites and notes that regions of the cortex project directly to the visceral periphera.

Moral disgust pervades the discourse surrounding transgressive figures such as Alex Rodriguez and Lance Armstrong. What particularly awakens our moral disgust is their rationalization of their behavior: in their minds, many others are doing the same thing, so they are just trying to keep up with the crowd. In their view, sport is a culture of lying, and that offends our innate sense of fairness and justice.

Disgust reactions provoke a desire to withdraw and a pervading sense of moral revulsion. Distress, conversely, provokes approach behaviors in many contexts. These responses are ancient and built into a biology tied to social dependence: we need one another. Social sports are one expression of these needs. So we are morally repulsed by cheating athletes.

But state-sponsored cheating may provide a legitimating context, encouraging a broad-based conception of getting ahead at all costs. We have evolved as a species that needs to live together in groups and maintain alliances. And the state, in certain circumstances, can become that group, that alliance.

Team- or college-sponsored sport can function in the same way, allowing transgressions to occur and be justified: the college president looks the other way since the team is winning; the team owner who wants home runs to get the fans back ignores what is going on in the locker room; a coach does not investigate the changing shapes and sizes of his athletes. The collective acceptance overrides the sense that the individual may have that the cheating is wrong. Cheating is not so hard to accept when couched in this legitimating collective context. Winning takes precedence over morals because the group has become the ultimate morality. We look away; we don't want to see.

Yet some people still protest. Sport is a crucial part of our human experience, and all the highs and lows of human expression are present in sport. Thus, embedded in sports is a view of ethics that goes only so far, for we are much more than the mere moral sentiments of Smith or Darwin or the pragmatic modification of Mill and the early utilitarians.

The deep-seated sense of disgust with cheaters is tied to one core sentiment: disgust at wrongdoing. Recall, for instance, Mark McGwire's 1998 season. Breaking the home-run record was exciting and thrilling, but underneath it, and finally emerging, was a visceral sense that what had happened was just not right, just not fair.

The larger sense is that moral culpability is social as well as personal. The ethics of work and individual initiative, the norms of moral discourse and social presentation, are embedded in the social milieu. It does not take much, unfortunately, to legitimize wrongdoing, but it also does not take much to make people look again and reevaluate behavior in that same social context.

Gut reactions are fast and often accurate appraisal systems. Problem solving in general, and with regard to morality in particular, is fast rather than reflective. It is ultimately fallible, and context matters. When it comes to making ethical decisions, we tend to favor rationality and denigrate emotions. But from an adaptive biological point of view, emotions are an integrated and regulated system that can facilitate goals. For instance, getting motivated requires emotion in all sports. And the control of expression, the degree of focus, and the capacity for control and direction are necessary in every endeavor. Emotions play a critical role in adaptation and regulation in accuracy, excellence, intelligence, endurance, and working through discomfort.

Jim Thorpe is considered one of the best all-round athletes America has ever produced (Buford 2010). A Native American, Thorpe attended the Carlyle Indian Industrial School in Pennsylvania, where he played football and ran track. At the 1912 Olympics he set pentathlon and decathlon records that held up for decades, but he lost his medals the next year when it was discovered that he had played baseball with a professional team for a few seasons. Thorpe returned to playing professional sports, both baseball and football, until 1929. He excelled in sports despite an adverse social context, and most believe that the loss of his Olympic medals was deeply unfair.

Avery Brundage also competed in the 1912 Olympics. He lost to Thorpe in the decathlon and pentathlon, but having remained technically an amateur, he went on to win a number of national championships. He later founded his own construction company and became a sports administrator. In 1936, Brundage fought the proposed boycott of the 1936 Olympics, which was held in Germany. Although Brundage's battle to take an American team to Nazi Germany was controversial, he was elected to the International Olympic Committee (IOC) that year and became IOC president in 1952 (Guttmann 1984, 1992, Mandell 1984).

Thorpe struggled most of his life. Brundage became an international leader. Fairness is frail, with pockets of clarity amid the omnipresent human tragedy in life and in sport.

What is fairness? It is linked to our notion of rights, and in the West our notion of rights is tied to our Enlightenment sensibility and a long history of events. The Magna Carta of 1215 defined the right to a fair trial and equality before the law. What would eventually become the English Bill of Rights during Britain's Glorious Revolution (1689) established the concept of citizens' rights within a polity. As we separated from religious authority, rights came to include individual choice of belief divorced from government. The Universal Declaration of the Rights of Man following the French Revolution in 1789 expressed universal rights for all and stressed the link to thinking for oneself (Kant 1787, 1792). Around the same time as the Universal Declaration of the Rights of Man, across the Atlantic we saw the American Declaration of Independence in 1776, Constitution in 1787, and Bill of Rights in 1791. And the concept of rights has continued to expand. In the United States, important milestones include the abolition of slavery (1865), the right to vote for African American men (1866), women's suffrage (1920), second-wave feminism and the civil rights movement (1960s), and the recent Supreme Court decision in the battle for marriage equality for homosexuals. But the concept of rights is never static. In an international context, the American emphasis on political and civil rights is criticized for its neglect of economic and social rights.

Women have been involved in sports for a long time. Spartan women are well known in the context of competition, and the Romans had women gladiators. Ancient Egyptian women competed in sports as well, and in some ancient African cultures women and men competed together in a number of games (Guttmann 1978, 2004).

But modern Western culture has only recently welcomed women's participation in sport. In the first modern Olympic games in 1896, there were 295 male athletes and no female competitors. By 1948, the games had expanded to include over 3,500 male athletes but fewer than four hundred female athletes. By the 2000 Olympic Games, however, the numbers had become more even: 6,582 men and 4,069 women (Guttmann 1991).

The passage of Title IX of the Education Amendments of 1972 was a landmark in U.S. women's sports. It states: "No person in the United States shall, on the basis of sex, be excluded from participation in, be denied the benefits of, or be subjected to discrimination under any edu-cation program or activity receiving federal financial assistance." Since most national championships are filtered through varsity programs and because many professional athletes come out of college teams, Title IX greatly facilitated women's participation.

In 1971, before Title IX was passed, there were around 172,000 male intercollegiate athletes and just under 32,000 female intercollegiate ath-letes. However, in 1986, after the adoption of Title IX, the number of female intercollegiate athletes increased to 83,000, whereas there were still nearly 172,000 male athletes (Guttmann 1991).

The current struggle for women in sport is being waged in Muslim countries, where religious requirements and cultural norms of modesty have made it difficult for girls to train, even in private, let alone compete in public. An example of this is the contention that arose among the more conservative Muslim leaders after Farah Ann Abdul Hadi, a Malaysian gymnast, wore the typical leotard at the 2015 Southeast Asia Games (Sang-hani 2015). Regardless of the controversy, Hadi proudly won six medals at the games for herself and her country. Social pressure can push both ways, however; the national desire to win Olympic medals is trumping the impulse to keep women out of the limelight. For example, Iran sent eight women to the London Olympics in 2012.

FIGURE 8.3

Fairness and gender.

Sport is never free of unfairness. Women were not allowed to compete in Olympic ski jumping before 2014, a sport where physiologically they may have an edge. In 2005 Gian Franco Kasper, a member of the IOC, said the sport "seems not to be appropriate for ladies from a medical point of view." But in 2014, women competed at the Sochi Olympics in ski jumping, with no injuries or obvious detriments to their health.

The disparity in award money, promotion, and sponsorships between the U.S. women's and men's national soccer teams became public knowledge after the U.S. women won the 2015 World cup over Japan in a 5–2 victory (Harwell 2015). Even though the Women's World Cup championship match was one of the most popular televised soccer games in American history, including the 2014 Men's World Cup in Brazil, the U.S. women's award for their hard work and perseverance was a mere $2 million; the champions of the Men's World Cup in 2014 won $35 million. Furthermore, the unfairness demonstrated by FIFA in their treatment of the women's teams became even more evident when the women players left the turf fields with painful burns because the turf was reportedly 120° Fahrenheit (Payne 2015). The Men's World Cup was played on standard grass fields and resulted in no burn-related injuries.

But sport often leads the way in the expansion of rights and freedoms. One has only to think of Jackie Robinson, depicted in Figure 8.4, breaking the color barrier in baseball (Rampersad 1998). The talent in the old Negro baseball leagues was phenomenal, but it took Branch Rickey, the owner of the Brooklyn Dodgers, to give Robinson the opportunity, and it took an athlete of Robinson's character and talent to take that opportunity and make the most of it.

In 2014, Derrick Gordon of the University of Massachusetts became the first openly gay collegiate basketball player, and Michael Sam became the first openly gay player to be drafted into the National Football League. Russia's homophobic laws targeting "the proselytization of children" brought out droves of gay athletes at the Sochi Olympics. Sports, long a testosterone-rich preserve of manliness, seems to be opening up to the concept of homosexual athletes.

Freedom is a key theme, but freedom does not function in a vacuum— it exists within our life context. Freedom under the law for the wealthy is not the same as that for the poor and the less privileged. While

FIGURE 8.4

Jackie Robinson was the first African American to play Major League Baseball.

equality under the law is a prime value, it is hard to accomplish. Sport is rife with the privileged and the less privileged. Freedom of opportunity is crucial.

Autonomy is important in moral responsibility, but we are also tied to institutions, groups, and states. Autonomy is mostly about thinking for oneself and when seeing that a wrong is being committed being willing to address it. But social context can make this more or less difficult to do. Social hope is an unhelpful social context's counterbalance.

SOCIAL HOPE AND SPORT

Social hope requires transactions that embrace the differences among us, a conception of the family of humanity—a humanism emboldened by natural piety and a broad common faith based on democratic participatory sensibilities (Dewey 1908, 1910, 1925). Social hope is the stuff that ties us together. Our evolution is bound by our social contact and by our sense of and concern for one another (Rorty 2000).

One lesson of our species' history is that we are frail and endlessly labile, promiscuous and bursting with possibilities, with a glorious sense of being connected to others, in participatory labors of self-initiative, self-excellence, and self-preservation. Social hope is naive, perhaps, but it is emboldened with possibilities and rife with contradictions. Social hope requires the intelligence to be anchored in, but not frozen to, things that matter. An expansion of participation through embodying this process (to varying degrees) underlies the conception of a participatory democracy. So does participation in social groups toward rational ends, where rationality is tied to the clarification of issues and the adjudication and the ability to compromise in context without devolution of principle and purpose.

Alexis de Tocqueville (1848) witnessed at first hand the growing sense of rights and social bonding, of "a right of association" and "acting in common," that was part of the origins of modern sport. In the middle of the nineteenth century, he documented how common bonds of labor and meaning were at the heart of freedom of expression and social bonds of value. The celebration was about democracy within self-initiatives coexisting with social hope—a social hope chosen, not imposed. Social hope, always frail yet omnipresent in an evolving cultural milieu, must also be linked to maximizing freedom of expression. Sport is an ideal of this cultural trend.

Social hope is the ideal of balancing in context both individual freedom and social bonds and responsibility, with an eye for inclusiveness as our cephalic capabilities expand. There is no panacea here, there are no miracles, just hard slogging amid a lot of intelligence and social practices. We need to acknowledge diverse forms of Thanatos, or devolution of function, as well as Fortuna, or luck, while being mindful of our natural continuity with others (Dewey 1908, 1925). The naturalistic urge to forge consequences lies in the participation in group formation, which is fraught with conflict though tempered by a democratic sensibility, which includes the participation of the least well off. Sport offers such opportunities.

It is a core cephalic adaptation to forge links through problem solving. Intelligence raises us above problem solving toward a rationality based not on deduction from absolute premises but on engagement with others, finding ways to forge ahead. Rationality toward the higher ideals, a "common faith" (Dewey 1934b) tied to a "common good" amid an evolving sense of rights, becomes a common expression, an expansion of human dignity, frail and fraught with endless disappointments. These are the alluring ideals that beckon to us, and they are made manifest in our sports.

Our transgressions match and complement our progress. Human beings are demythologized on this journey. And it is a journey not toward perfection but toward peaceful and respectful coexistence—yet endlessly marked by real danger, real wars, and real Holocausts. Worry remains our common currency, an ontic fact and a reason further to forge collaborative bonds, linking social hope to memory and forward-looking sensibilities.

Sports exist in the context of social hope and broader participation in a common context of human expression, striving to release the "better angels of our nature" (Lincoln 1861, Pinker 2011). President Mandela of South Africa, an avid sports fan and a boxer himself, managed to use the all-white South African rugby team to unite a country recovering from the trauma of apartheid. He was a true practitioner of social hope and one of our better angels.

CONCLUSION

Fairness and ethics are built into the fabric of our activities. Sport is just one good example. We need rules for sport and for all social activities. When the rules are not fair, we look to change them. We admire the character of people who follow them fairly; we do not admire people who cheat.

Of course, character pervades sport; showing character even when one can do nothing about a bad call, persevering despite it, showing one's best side—we applaud these features. Getting up and carrying on despite adversity, showing character under stress: these fuel comebacks, and they are normative goals in sports and in the rest of life.

9

DIGNITY AND BEAUTY

I do not much believe in definitions, and human dignity is a concept that is particularly hard to define. It is largely described by a range of relationships (Wittgenstein 1953). There may indeed be something like an implicit concept of dignity, but it is hard to capture in words. Like Kant's view on aesthetic judgment, it does not fit easily into a bounded concept (Kant 1787, 1792). Beauty runs through dignity with tapestries of awe-inspiring elegance.

In spite of these challenges in defining dignity, when we watch individuals or groups persevere through hard times and we value their intended goal, we understand the sense of dignity: dignity in death, under hardship, amid defeat, or when winning. Dignity is tied to fairness. Sport is an ideal medium in which to express dignity, and it is a significant part of sport's appeal.

Professional or amateur athletes trying to play in the face of injury or training to get back from injuries are features of many heroic sports stories. I remember watching the basketball player Willis Reed emerge, battered and hurt, for the final competitive battle in a critical game against Los Angeles. Reed, a short center at maybe six feet, eight inches, had a beautiful jump shot and was a fearless competitor. Sport is reality for the better part of our nature.

The Willis Reed story is a commonplace one in sport, showing athletes competing under difficult personal or professional conditions when it counts. When we think of dignity in sport, we think of Kirk Gibson's home run, hobbling across the bases to home plate to win the game for the Los Angeles Dodgers against their rival, Oakland, in the last inning of the 1988 World Series. Or we think of Kerri Strug continuing to vault on an injured ankle and being carried to the podium to accept her team gold medal by her coach Béla Károlyi at the 1996 Olympics. Events like these inspire us and keep fan loyalty and fan identity alive. It is the connection to athletes in these emotive moments that keeps fans coming back time and time again even if, as I understand as a long-time New York Knickerbocker fan, their team doesn't win much.

Coaching is central to many of these moving moments in sports history. Comebacks depend on genes, capability, and discipline, but coaching is also vital. Sport occurs mostly in a social context, but the individual is often at the heart of the performance of the social sport. Coaches make or break athletes by honing their performance, execution, and goals. The art of coaching lies in a context of "meaning" through interactions with the athlete (Stelter 2007). The coach determines the manner of instruction, and then the athlete adheres or does not. The facilitation of motivation is one interaction that is expressed throughout sport and coaching, and it is essential under most circumstances.

This facilitation of motivation can happen in a wide variety of ways, but it is possible to describe what often might work best in coaching from both the coach's and the coachee's point of view. Successful coaches listen well, offer good action-point ideas, state clear objectives, have no personal agendas, are accessible and available, provide straight feedback, are good models of effectiveness, and have seen other career paths. From the coachee's point of view, a successful coach is generally one who reflects, cares, continues to learn, checks back, is committed to the coachee's success, demonstrates integrity, is open and honest, knows the "unwritten rules," and pushes the coachee when necessary (Passmore and Gibbes 2007).

Memory does not fade as I envision Phil Jackson, sixth man for the Knickerbockers, during the period when Willis Reed, Walt Frazier, Earl Monroe, Bill Bradley, Dave DeBusschere, and Jerry Lucas played—this was so many years ago in 1973, but to this fan their names read off as if it were yesterday. They were world champs and a formidable team, and the great coach Red Holzman was the key to their success. He stressed sharing the ball and looking for the open person who had the best shot. A group of players working so well as a team is a thing of beauty to watch. Phil Jackson went on to become one of the finest professional basketball coaches of all time. He was, of course, fortunate to have the likes of Michael Jordan, Kobe Bryant, and Shaq as his mentees. We recall the endless array of wonderful coaches across time.

We find inspiration for the meaning of life in sport: dignity, social contact, rising to show the "better angel" overcoming adversity, managing defeat, the wondrous sense of well-earned and arduous victory, graciousness toward others in their defeat. Yet we also have the other side of events. Remember the ice skaters Tonya Harding and Nancy Kerrigan. This is the vile underbelly of sport, and there is plenty of it. Take, for instance, competitive parents who end up hurting one another or their children. There are all too many instances of the bestial within the elegant in the human condition. Sports display all the aspects of Shakespearian drama in an up-close and personal setting.

SPECIAL OLYMPICS

But we still have a phenomenal capacity to reach out to others, as Adam Smith noted when he discussed the moral sentiments (1759). This is explained as a capability in our social nature (Darwin 1859, 1871): something nurtured in a social context where it can materialize in a culture where consideration and action are understood within a context of fairness (Rawls 1971). Those who have less can elicit the better angels; one place where this is possible is the Special Olympics. The Special Olympics became a reality in 1963 in the United States; the international variant began some five years later.

When I first came to Washington I interacted with Dr. Cooke, who was close to the Kennedys. He played a role with Rosemary, John F. Kennedy's sister, who had had brain surgery to reduce her aberrant behaviors (with some unfortunate results). Dr. Cooke was close to Eunice Shriver, Rosemary's sister, and Eunice's husband, Sargent. The two of them, with the aid of many others, ignited what is now the Special Olympics. This competitive event features athletes with special needs. Shriver's experiences with her sister may have inspired her desire to create these events.

What is special about Special Olympics is that it gives individuals with intellectual disabilities a chance to compete, when they might not

FIGURE 9.1

Eunice Kennedy Shriver, an advocate for sports education and activities for children
with special needs.

Source: Courtesy of Special Olympics.

ordinarily have that opportunity. Competition, a basic drive of our species, is important in creating and sustaining human dignity. Enabling
individuals to experience competition on a large scale helps give and
make meaning, and it is an adaptation fundamental to our mental health
(Jaspers 1913). These games are held within a special context, celebrating
those among us with disabilities. At its best, it is about human meaning
and performance and striving to be better and get better. It is a celebration
of human meaning amid struggle and adversity.

From alpine skiing to volleyball, basketball to sailing, and judo to
powerlifting, the types and range of sports in the Special Olympics are
rich, and the competition's growth has been outstanding. Meanwhile,
the impressive number of sports in the Paralympic Games has ultimately facilitated and reinforced the improvement of opinions and atti-
tudes toward individuals with disabilities, particularly athletes (Gold
and Gold 2007). Boccia, five- and seven-a-side football, and wheelchair

FIGURE 9.2

Special Olympics.

Source: Courtesy of Special Olympics.

versions of basketball, tennis, curling, and fencing are all part of the Paralympic Games program.

Human dignity and solidarity are expressed in the context of the Special Olympics, and adaptation, well-being, and the role of sport are important elements in the context of sport. These are as special in the Special Olympics as they are in peewee-league baseball or NFL football. Training under these diverse conditions requires excellence of both the coach and the Olympian, impaired or not.

Amid the biology of hope and the endless intertwining of cultural evolution with greater participation in the making of meaning, the Special Olympics stands out as a shining example of the service that sport can do for the world. The way we treat and support the disadvantaged, within the bounds of reason, is a measure of our worth.

THE DIGNITY OF THE UNDERDOG

The opposite side of eugenics is that our "better angel" often roots for the underdog: the athletes and teams who have to work harder than others, who rise up under conditions of adversity and are not expected to win, whether it's the United States versus Russia in 1980s hockey, amateurs versus professionals, or Rocky Balboa.

Conversely, many people hated the New York Yankees' "buying" wins by acquiring players using the enormous resources at their disposal. Of course, money and players bought does not necessarily add up to winning, although it may buy having an edge in the contest. But the mindset is inherently unfair and unsportsmanlike. For many Americans, "those damn Yankees" were about as far from the underdog as you could get. In the Broadway musical *Damn Yankees*, an elderly man was willing to sell his soul to the devil to join the losing Washington Senators and, as the ultimate underdog, defeat the Yankees.

The underdog grabs our hearts, and heart is a big part of winning— especially winning when it really matters. No surprise, then, that one song in *Damn Yankees* is "You Gotta Have Heart." So we often root for the underdog—except if the underdog is playing against our team. Our social sense is seeing those with less get more; it is one motivation among many, and perhaps a frail one. But sport reveals it. Rooting for the underdog goes all the way back to the Bible; think about David and Goliath or the blinded Samson. Perhaps this is in our nature; after all, it is an expression of an "us" in which biology and culture converge in an expression of rooting for those with less.

Cultural progress is about broad participation in opportunity, and sport is a primary example of this. We do not leave biology and culture on separate sides: they run into each other, like theory and evidence, like fact and value (Dewey 1908, 1925). The revolt of dualisms is evident in sport; cephalic capabilities and predilections merge with training and sport options. Appraisals are rich in values, and values are inherent in our facts. And these facts are not mere social constructions; they are real, and sport endlessly reminds us that they are.

Perhaps our cheering for the underdog stems from our knowledge that evolutionary change is far from linear. There is no straight arrow forward, just adaptive radiation, extinction, and diverse forms of stability and breakdown in equilibrium (Gould 1977). The underdog may become top dog in the right circumstances, and circumstances are always changing. Both evolution and devolution of function are common, perhaps particularly in our age, since we live longer and care for those who would not have survived in previous times in our history as a species.

Sport is thus a wonderful arena for us to showcase human dignity. Consider, for instance, the case of Magic Johnson, a truly magical basketball player who was diagnosed with HIV, a much-stigmatized and feared disease in the 1990s. Johnson changed the public perception of those living with HIV by his admission of his illness, his remaining a public figure, and his friendship with Larry Bird—or rather, the evolution of that friendship, which grew stronger in the context of that disease. When some did not want to touch him or play with him, Bird and many others embraced Magic Johnson. Fierce competitors find friendship enhanced by human frailty, human meaning, and social contact (Jaspers 1913).

Jim Abbot, a baseball player for the Yankees, played in the big league as a no-hitter. He wanted to be known as a great pitcher and not for the fact that he is missing a hand. But he was and is an inspiration in his sportsmanship and his humanity (Abbott and Brown 2012).

FIGURE 9.3

Magic Johnson and Larry Bird.

Source: AP Images.

COEVOLUTION OF ANIMALS AND HUMANS

Domestication, which predates sport, involves both genetic and epigenetic adaptation in animals. Domestication of the dog is surprisingly ancient, going back possibly more than thirty thousand years in Europe, for example (dogs were likely independently domesticated in several different parts of the world) (Yong 2013).

Canine domestication has affected a broad range of human activities. Dogs became a companion to humans not only in recreational activities and predation practices but also in competitive sports, such as dog racing and dog sledding. Sport and domestication are as linked as hunting is to our cultural evolution.

The same holds true for horses. As they do with dogs, humans form social connections with horses. Their domestication has played a significant role in human society, especially among pastoralist societies that relied on horses as a means of transport. Eventually, horses transitioned from playing only a functional role in human life to also providing entertainment via sport.

Consider the range of ways in which we use horses in our sports. There are four equestrian sports in the Olympics: dressage, eventing, show jumping, and classical dressage. Steeplechase, harness racing, and thoroughbred horse racing are the three main categories of racing. Also, English riding and Western riding are distinguished by not only their riding style and saddle but by the differing kinds of sports events for each. For example, classical dressage, polo, and fox hunting are performed by English riders whereas, reining, calf roping, barrel racing, and chuckwagon racing are performed by Western riders.

The domestication of the horse and its adaptation from work and from war to sport reflects our cultural evolution: our cognitive and physical capabilities are expanded by the use of the horse.

Horses and dogs have evolved to please human beings. Cats are a different story. What is interesting is which species can and which cannot be domesticated to participate in a culture of sports and competition. Though cats engage in active play and have been domesticated since at least the time of ancient Egypt, probably first drawn to human agricultural settlements because of the rodents attracted to grain-storage facilities

FIGURE 9.4

An early domesticated horse.

(Diamond 1997), they are not as willing as horses and dogs to please their human owners. It is their desire to please that humans exploit in getting animals to participate in competitive sports—not any innate desire in these animals to compete formally. Sport is a human behavior and a human predilection.

For horses and dogs, epigenetic changes are taking place; genetic material is either being expressed or not, and perhaps the effects of our joint evolution is revealed in both. Sport is just one place in which it is expressed. One feature in both humans and our animal sporting partners may be the expression of oxytocin in regions of the brain; that chemical is implicated in the maintenance and enhancement of the social bonds between us in cooperative social contexts (including sports).

SPORT, AESTHETICS, AND BEAUTY

In *Art as Experience*, John Dewey made it clear that aesthetics is a fundamental feature of human experience: "Art is prefigured in the very processes of living" (Dewey 1934, Johnson 1987). Art and aesthetics are built into our adaptive capability, in our building and nesting and attaching to others. Dewey's scholarship reveals the continuity of biology and culture.

Aesthetics is not just something exclusive for the galleries, concert halls, and performance spaces. Aesthetics lies within the biology that we bring to our world, within the culture that we live in or are adapting to. We naturally pay attention to the contours of shape and form, manifesting our innate geometrical capabilities. That is one reason that we are in awe of the magnificence and sublime features of nature (Kant 1792); nature is within our grasp at certain moments but then lost in the vast space of beauty and power.

Sport is rich in awe-inspiring expression. We marvel at the beauty of inspiring turns, balletic movement, and the match of music and dance as ice skaters swerve and leap on the ice, individually or with a partner. And the perception of beauty and biology may not be exclusive to our species. Beauty is often tied to fitness, and sport is most definitely about fitness, which is knotted to our attention. When Sarah, an adult research chimpanzee, was shown different dancers, she most attended to the one that her human trainers agreed showed the more beautiful form in the dance (Premack and Premack 1995).

FIGURE 9.5

Figure skaters representing the balletic quality of their movements.

The link between form and function, so fundamental in aesthetics, ranges across the full spectrum of cephalic function; ballet, for instance, is invoked as a comparison across many forms of sports. Getting there requires thought; as Dewey (1934) asserted, "An experience of thinking has its own aesthetic quality." And thinking is richly expressed across sport, across training, across persevering. No separation of a mind in a body, just the blending into coherent action, anticipatory control.

What pervades sport and aesthetics is an appetitive and consummatory experience: the stuff that underlies action and motivation. The appetitive side is the desire manifest in action, which is followed by the consummation of some sort of satisfaction. Such experiences are common in biology and cephalic capabilities (Dewey 1925).

The practice of sport, as well as its observance, is an aesthetic experience. Athletes work hard, practice after practice, their activity rich in appetitive and then sporadic consummatory experiences: motivation and satisfaction, pain, redundancy. Sport is not easy, but the aesthetic rewards are strong. We often say that top sports figures are driven to excel, and, indeed, without drive not much emerges. But techniques that promote peacefulness are also essential for sport (for example, mediation, tai chi, and mindfulness techniques that can enhance attention) and for life more generally, perhaps in some sports more than others. Though some great athletes make it look easy, drive wins out as long as there is excellence at the base and good work habits throughout.

So what motivates us in sport? One answer is the enrichment and intensity of the experience, in the same way that good art enriches and intensifies experience. What motivates is the enhancement of experience. Sport brings a confluence of cephalic capabilities in form, function, and achievement. We go beyond the sensory whenever we pursue pleasure; it is not a simple hedonic calculation, though sensory pleasure is a core feature in sport. Achievement and satisfaction are also derived from meeting goals and expectations.

Then, of course, sport plays a key role in the rituals of life; rules serve that role as we ascend in the continuation of the sport. Bouts of completeness emerge across the contours of the sport experiences. The biological rhythms that permeate cephalic activity also permeate our cultural side and the diverse organ systems in the body (liver, adrenals, etc.). Indeed, we are in tune with the rhythmic movements of our planet and our local niche; the measurement of time is a feature of the brain in our adaptation to the world around us (Richter 1979).

Much of sport is about rhythm: in swimming, basketball, hockey, running—almost every sport. Like most of life, sport demands order and novelty, twin features of rhythm. Expectations oscillate, along with the emergence of the novel and unexpected, the give and take of continuity and change; this is the stuff of heightened experience and rich adaptation.

Aesthetic experiences permeate sport, as our experiences are heightened (Dewey 1925, 1934). This sense of heightened sensibility is apparent across the array; consider the horse sports. We have the majesty of the horse itself, our cultural selection of characteristics that we admire and that are useful to us in work, in hunting, in war, and in racing. Aesthetics are built in the relationship between us and horses, and we coevolved as we traversed the familiar to the unfamiliar. The relationship between horse and human is also tied to our representation, our understanding, our creations, and our explorations. Creating and understanding are often intimate, and such relationships pervade sport, domestication, and the search for excellence.

FIGURE 9.6

George Bellows, *Stag at Sharkey's* (1909).

Aesthetics runs through the whole of human experience; that is why Dewey called it "art as experience." Sport is rich in aesthetics, rife with the perfection of the body, the sense of excellence appropriated from the Greeks and the founders of the Olympic competition. Bodily perfection, training, and performance are diversely expressed across cultures; the array is endlessly rich—an anthropology of the human condition. Art and sport have a conjoined history. From the classical period onward, a rich tapestry of expression reveals the connection between aesthetics and sport. Art loves sport because beauty and form and excellence are apparent in it.

Boxing, for example, is a brutal pastime, but it can be rich in beauty. George Bellows was a master of demonstrating its elegance and its coarseness, its stimulation and its casual violence. His paintings illuminate, at a fundamental level, the connection between aesthetics and sport: even in the most brutal of sports.

SPORT AND SUCCESSFUL AGING

S port is a lifelong activity, and for some people, that is very long indeed. The late Oliver Sacks (2015), who was a well-known neurologist, described his father and himself in the water. His sense of well-being while swimming is tied to being physical, but it was also tied to a sense of being with his father. He also describes swimming as being in a place where he is not reachable in this wired age. For Sacks and his father, it was a respite from the travails of life, a respite into the physical and into practice, a mind in a body, a sense of well-being. This is something we should learn early on and an experience that remains important throughout life.

Sport is anchored to diverse means of generating health through physical activities and mental games. Physical action may not generate happiness on its own, but it sure helps. We move from "winning at all costs" and being vulnerable to injury to sustaining what we have and cultivating health in the context of movement and action.

It is nontrivial that being successful involves taking care of cephalic systems (the mind-body continuum); aging successfully involves doing with less and making the most of it. In fact, achieving excellence in sport is a process of maximizing and taking advantage of resources and then sustaining them. Of course, these features underlie a good deal of adaptive behavior.

Aging can be hard to handle for the athlete. Athletes are wired to win, to push hard. Sometimes ex-athletes push too hard, finding it hard to acknowledge that their cephalic systems are not what they used to be. But the endgame is a reflex of what once was; what is hard about aging is not

FIGURE 10.1

Oliver Sacks.

Source: Marsha Grace Williams.

the old body but the new one; athletes must remember that, in order to preserve what they have and avoid injuries.

And what is clear is the richness of the deep relationship between biology and culture; sport makes full use of what makes us biologically human and demonstrates that our cultural evolution has expanded those potentials.

And then within this view are the tools evolved for instrumental exploration across sport, linked in the efficiency and adaptation that figure across our exploration of neural function. Design principles are efficiencies that reflect our species and its capabilities for survival in local niches. The neural systems reflect specificity, separation, minimization. "All things in light of evolution" (Dobzhansky 1962), and, it is always apparent, all things in light of culture. Culture occurs endlessly within an evolutionary context (Boyd and Richerson 1988), and events like the capability to throw accurately figured importantly in our evolution (Calvin 1982). Sport makes this transparent; biological and cultural continuity are palpable in sports.

Designer swimsuits, skis, bats, sleds, boats, paddles, etc., and training and practice schedules all exist to achieve better performance.

Sport success is a combination of tool making, an expansion of ourselves in practice, the talent that we bring to the table, and the mindset
to win. Cephalic capability, coupled with culture, learning, training,
and sheer fortune, is what allows the expression for sport and sense of
solidarity with sport. Sports nationalism, along with sporting solidarity, coinhabits the human space of meaning (Dewey 1934).

There is no one theory about sport or one definition of it (Baker 1982,
Guttmann 1978, Mandell 1984). Across the definitions are important similarities and many differences. What pervade are normative notions of fairness, civility as an ideal, and the importance of dignity and growth. Sport
is—ideally—civilizing, or it can be, and it represents values that many of
us hold dear: fairness, hard work, and an evolving ethic of participation.

And there is dignity. Some individuals do a lot with a less-than-ideal
start. We have a special sort of Olympics where sport is inclusive, and
inclusiveness exemplifies our expanding sense of participation as human
expression. And sport brings us together—yet it also divides because of
nationalism. As in all things human, there is the good, the less than good,

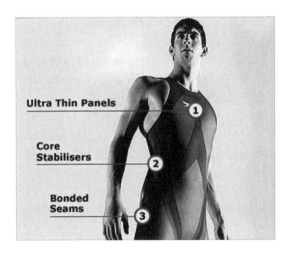

FIGURE 10.2

LZR Racer suit.
Michael Phelps posing in an advertisement for the LZR Racer suit.

Source: http://news.bbc.co.uk/sport2/hi/olympics/swimming/7328469.stm.

and the ugly. The pursuit of excellence, as hard as that is, is pervasive in sport. The valuation of excellence, motivation, and sport achievement reaches deep in our shared human experience across cultures. It is embedded in our biology.

The building block for sport has roots outside of sport; sport is part of culture. And play, though found in many species, is particularly pervasive in ours—our behavior merges from play to sport, given a suitable cultural context. Pedagogy and getting a foothold in the world of others, playing with others, are rooted in this world, in which we depend on one another and are constantly adapting. Sport is one vehicle for this.

Evolutionary steps led us to become the sporting species that we are: notably, an erect stance, an expanded shoulder capability for throwing the tools we developed to catch prey and to fight, and a developed Achilles tendon and the resulting ability for long-distance hunting—tracking animals over time and space—and then sustaining the running behaviors in social-cooperative manners. And not surprisingly, the size of an expanded cortex is tied to the degree of social complexity (Dunbar 1992).

An evolutionary perspective is one route to understanding sport; blended with our biology, culture fuels our capabilities. The evolution of the brain and other organ systems were vital steps in how we became able to do sport and how other hominids did not. And research has now entered the arena of genetics and epigenetics in understanding sports capability (metabolism, oxygen consumption, capacity for running, etc.).

And running itself promotes neurogenesis in regions of the brain, including the hippocampus; memory capability is essential for learning, and learning is tied to the long pedagogy that goes from play to discipline and then back to wonder (Whitehead 1929, 1938).

The discipline of sport, the endless practices, are set in the larger context of striving for excellence in competition. And within this context are training to improve performance, the tools that make that improvement easier, and, of course, the cheating and broken hearts.

Though sport is about more than winning, winning is nontrivial. And there is a fine line between what is fair and what is not; what we want is transparency, and what is allowed for one person should be available for another. But life is not fair; some of us have resources, and a lot do not. But biology is also not fair: some people are amazing; most are not. Some people utilize their capabilities to a great degree; most do not. Some people avoid injuries, and some recover quickly, others less so. Biology matters crucially, including the biology of temperament and learning capabilities.

Most of us have enough of the right stuff to participate in diverse sports. This involves us in a form of enculturation and participation, team belonging, striving to participate, and the motivational allure to succeed in the context of the development of a social self.

While pedagogy is with us for a long time, so is aging, as is participation in sport and other forms of social contact. Teams and training and camaraderie: all these push up the motivational allure. Such exercise later in life has two effects that we have recently realized: it promotes neurogenesis and enhances memory capacity.

As we live longer, these are nontrivial effects. Add to them the secretion of brain endorphins. It is not hard to see sport's fundamental link to our sense of well-being. In practice and in performance, it reveals the utter integration of a mind in a body. This is one reason why sport is so revelatory. Another is that sport is also about beauty. And this is also nontrivial because fitness and adaptation are key to survival. This is not beauty in the abstract; it is beauty in performance, in action: the allure of sport from the audience's point of view and the aura of sport within athletic sensibility.

Of course, always present are the good, the bad, and the ugly: the cheating, the bad behaviors of coaches and teammates, the audience, and the competition. Sport cultures evolve within a biological framework; culture shapes some of the biology that is expressed in sport, and sport provides a telling window into the human condition. After all, it puts everything about us on display.

And in many parts of the globe, sport is transforming into a lifelong endeavor. Consider, for example, the older folks running marathons and the master softball players. I rejoice to see sport integrated into a way of life, a lifelong way of being.

REFERENCES

Abbott, J., and T. Brown. 2012. *Imperfect: An Improbable Life*. New York: Random House.

Abreu, A., et al. 2012. "Action Anticipation Beyond the Action Observation Network: A Functional Magnetic Resonance Imaging Study in Expert Basketball Players." *European Journal of Neuroscience* 35:1646–1654.

Adkins, D., et al. 2006. "Motor Training Induces Experience-Specific Patterns of Plasticity Across Motor Cortex and Spinal Cord." *Journal of Applied Physiology* 101:1776–1782.

Adkins-Regan, E. 2005. *Hormones and Animal Social Behavior*. Princeton, N.J.: Princeton University Press.

Adolphs, R. 1999. "The Human Amygdala and Emotion." *Neuroscientist* 5:125–137.

Adolphs, R. 2001. "The Neurobiology of Social Cognition." *Current Opinion in Neurobiology* 11:231–239.

Aggleton, J. 1992. *The Amygdala: Neurobiological Aspects of Emotion, Memory, and Mental Dysfunction*. New York: Wiley.

Aggleton, J. 2000. *The Amygdala*. Oxford: Oxford University Press.

Aglgioti, S., et al. 2008. "Action Anticipation and Motor Resonance in Elite Basketball Players." *Nature Neuroscience* 11:1109–1116.

Ajemian, R., et al. 2013. "A Theory for How Sensorimotor Skills Are Learned and Retained in Noisy and Nonstationary Neural Circuits." *Proceedings of the National Academy of Sciences*: E5078–E5087.

Aldridge, J. W., K. C. Berridge, and A. R. Rosen. 2004. "Basal Ganglia Neural Mechanisms of Natural Movement Sequences." *Canadian Journal of Physiological Pharmacology* 82:732–739.

Alembert, J. 1751. *Preliminary Discourse to the Encyclopedia of Diderot*. New York: Library of Liberal Arts, 1963.

Alheid, G. F., and L. Heimer. 1988. "New Perspectives in Basal Forebrain Organization of Special Relevance for Neuropsychiatric Disorders: The Striatopallidal, Amygdaloid, and Corticopetal Components of Substantia Innominata." *Neuroscience* 27:1–39.

Allen, J. S. 2009. *Human Evolution and the Organ of Mind.* Cambridge, Mass.: Harvard University Press.

Alstermark, B., and T. Isa. 2012. "Circuits for Skilled Reaching and Grasping." *Annual Review of Neuroscience* 35:559–578.

Altman, J. 1966. "Autoradiographic and Histological Studies of Postnatal Neurogenesis." *Journal of Comparative Neurology* 124:431–474.

Alvarez-Buylla, A., M. Theelen, and F. Nottebohm. 1988. "Birth of Projection Neurons in the Higher Vocal Center of the Canary Forebrain Before, During, and After Song Learning." *Proceedings of the National Academy of Sciences Neurobiology* 85:8722–8726.

Amaral, D. G., et al. 1992. "Anatomical Organization of the Primate Amygdaloid Complex." In *The Amygdala: Neurobiological Aspects of Emotion, Memory, and Mental Dysfunction*, ed. J. P. Aggleton, 1–66. New York: Wiley-Liss, 2000.

Arendt, H. 1963. *Eichmann in Jerusalem.* New York: Viking.

Aristotle. 1962. *The Nicomachean Ethics.* New York: Macmillan.

Aristotle. 1968. *De Anima.* Trans. D. Hamylin. Oxford: Oxford University Press.

Aristotle. 2008. *Physics.* Trans. Robin Waterfield. Oxford: Oxford University Press.

Arnold, A. P. 2002. "Concepts of Genetic and Hormonal Induction of Vertebrate Sexual Differentiation in the Twentieth Century with Special Reference to the Brain." In *Hormones, Brain, and Behavior*, ed. D. W. Pfaff. New York: Elsevier.

Asfar, B., et al. 1999. "*Autralopithecus garhi*: A New Species of Early Hominid from Ethiopia." *Science* 284:629–634.

Ashe, A., and A. Rampersad. 1994. *Days of Grace.* New York: Ballantine.

Aston-Jones, G., and J. D. Cohen. 2005. "An Integrative Theory of the Locus-Coeruleus-Norepinephrine Function: Adaptive Gland and Optimal Performance." *Annual Review of Neuroscience* 28:403–450.

Atran, S. 1990. *Cognitive Foundations of Natural History.* New York: Cambridge University Press, 1996.

Atran, S., D. L. Medin, and N. O. Ross. 2005. "The Cultural Mind." *Psychological Review* 112:744–766.

Babiloni, C., et al. 2010. "Resting State Cortical Rhythms in Athletes: A High-Resolution EEG Study." *Brain Research Bulletin* 81:149–156.

Baker, W. 1982. *Sports and the Modern World.* Urbana: University of Illinois Press.

Baldwin, J. D., and J. I. Baldwin. 1974. "Exploration and Social Play in Squirrel Monkeys (Saimiri)." *American Zoologist* 14:303–315.

Ball, G. F., and J. Balthazart. 2002. "Neuroendocrine Mechanisms Regulating Reproductive Cycles and Reproductive Behavior in Birds." In *Hormones, Brain, and Behavior*, ed. D. W. Pfaff et al. New York: Academic Press.

Barger, N., L. Stefanacci, and K. Semendeferi. "A Comparative Volumetric Analysis of the Amygdaloid Complex and Basolateral Division in the Human and Ape Brain." *American Journal of Physical Anthropology* 134:392–403.

Barham, L., and P. Mitchell. 2008. *The First Africans*. Cambridge: Cambridge University Press.

Baron, J. 1988. *Thinking and Deciding*. Cambridge: Cambridge University Press, 2008.

Baron-Cohen, S. 1995. *Mindblindness*. Cambridge: MIT Press, 2000.

Baron-Cohen, S., H. Tager-Flushberg, and D. J. Cohen. 1993. *Understanding Other Minds*. Oxford: Oxford University Press, 2000.

Barrett, L., and P. Henzi. 2005. "The Social Nature of Primate Cognition." *Proceedings of the Royal Society B: Biological Sciences* 272:1865–1875.

Barrett, L., P. Henzi, and D. Rendall. 2007. "Social Brains, Simple Minds: Does Social Complexity Really Require Cognitive Complexity?" *Philosophical Transactions of the Royal Society of London B: Biological Sciences* 362:561–575.

Barrett, P., and P. Bateson. 1978. "The Development of Play in Cats." *Behaviour* 66:106–120.

Barron, A., E. Sovik, and J. Cornish. 2010. "The Roles of Dopamine and Related Compounds in Reward-Seeking Behavior Across Animal Phyla." *Frontiers in Behavioral Neuroscience* 4:163.

Barsalou, L. W. 2003. "Abstraction in Perceptual Symbol Systems." *Philosophical Transactions of the Royal Society of London B: Biological Sciences* 358:1177–1187.

Barton, R. A. 2004. "Binocularity and Brain Evolution in Primates." *Proceedings of the National Academy of Sciences* 101:10113–10115.

Barton, R. A. 2006. "Primate Brain Evolution: Integrating Comparative Neurophysiological and Ethological Data." *Evolutionary Anthropology* 15:224–236.

Barton, R. A., J. P. Aggleton, and R. Grenyer. 2003. "Evolutionary Coherence of the Mammalian Amygdala." *Proceedings of the Royal Society B: Biological Sciences* 270:539–543.

Bauman, S., et al. 2005. "A Network for Sensory-Motor Integration." *Annals of the New York Academy of Sciences* 1060:186–188.

Beach, F. 1945. "Current Concepts of Play in Animals." *American Naturalist* 79:523–541.

Beach, F. 1979. "Animal Models and Psychological Inference." In *Human Sexuality: A Comparative and Developmental Perspective*, ed. H. A. Katchadourian, 98–112. Berkeley: University of California Press.

Bechara, A. 2005. "Decision Making, Impulse Control, and Loss of Willpower to Resist Drugs: A Neurocognitive Perspective." *Nature Neuroscience* 8:1458–1463.

Begliomini, C., et al. 2014. "An Investigation of the Neural Circuits Underlying Reaching and Reach-to-Grasp Movements: From Planning to Execution." *Frontiers in Human Neuroscience* 8:676.

Begun, D. R. 2003. "Planet of the Apes." *Scientific American* 289, no. 2: 74–83.

Beilock, S., and S. Gonso. 2008. "Putting in the Mind Versus Putting on the Green: Expertise, Performance Time, and the Linking of Imagery and Action." *Quarterly Journal of Experimental Psychology (Hove)* 61:920–932.

Békésy, G. von. 1959. *Experiments in Hearing*. New York: McGraw-Hill.

Bekoff, M. 1995. "Play Signals as Punctuation: The Structure of Social Play in Canids." *Behaviour* 132:419–429.

Bekoff, M., and J. Byers. 1998. *Animal Play: Evolutionary, Comparative, and Ecological Perspectives.* Cambridge: Cambridge University Press.

Bell, C., and H. Suggs. 1998. "Using Sports to Strengthen Resiliency in Children. Training Heart." *Child & Adolescent Psychiatric Clinics of North America* 7, no. 4: 859–865.

Bentley, P. J. 1982. *Comparative Vertebrate Endocrinology.* Cambridge: Cambridge University Press.

Berlyne, D. E. 1970. "Novelty, Complexity, and Hedonic Value." *Perception and Psychophysics* 8:279–286.

Berlyne, D. E. 1954. "A Theory of Human Curiosity." *British Journal of Psychology* 45, no. 3: 180.

Bernard, C. 1865. *An Introduction to the Study of Experimental Medicine.* New York: Dover, 1957.

Berridge, K. C. 2004. "Motivation Concepts in Behavioral Neuroscience." *Physiology and Behavior* 81:179–209.

Berridge, K. C. 1996. "Food Reward: Brain Substrates of Wanting and Liking." *Neuroscience & Biobehavioral Reviews* 20:1–25.

Berridge, K. C., and T. E. Robinson. 1998. "What Is the Role of Dopamine in Reward: Hedonic Impact, Reward Learning, or Incentive Salience?" *Brain Research Reviews* 18:309–369.

Berthoz, A. 2002. *The Brain's Sense of Movement.* Cambridge, Mass.: Harvard University Press.

Bickart, K., et al. 2011. "Amygdala Volume and Social Network Size in Humans." *Nature Neuroscience* 14:163–164.

Binofski, F., and L. Buxbaum. 2013. "Two Action Systems in the Brain." *Brain and Language* 127:222–229.

Blackburn, E. 2005. "Telomeres and Telomerase: Their Mechanisms of Action and the Effects of Altering Their Functions." *FEBS Letters* 579:859–862.

Blake, E., and I. Cross. 2008. "Flint Tools as Portable Sound-Producing Objects in the Upper Paleolithic Context: An Experimental Study." *Experiencing Archaeology by Experiment*: 1–19.

Blakemore, S., and J. Decety. 2001. "From the Perception of Action to the Understanding of Intention." *Neuroscience* 2:561–567.

Blennow, K., J. Hardy, and H. Zetterberg. 2012. "The Neuropathology and Neurobiology of Traumatic Brain Injury." *Neuron* 76:886–899.

Boecker, H., et al. 2008. "The Runner's High: Opioidergic Mechanisms in the Human Brain." *Cerebral Cortex* 18:2523–2531.

Booth, F., M. Chakravarthy, and E. Spangenburg. 2002. "Exercise and Gene Expression: Physiological Regulation of the Human Genome Through Physical Activity." *Journal of Physiology* 543, no. 2: 399–411.

Bostan, A., R. Dum, and P. Strick. 2013. "Cerebellar Networks with the Cerebral Cortex and Basal Ganglia." *Trends in Cognitive Sciences* 17:241–254.

Bouchard, C. 2012. "Genomic Predictors of Trainability." *Experimental Physiology* 97:347–352.

Bouchard, C., et al. 2000. "Genomic Scan for Maximal Oxygen Uptake and Its Response to Training in the HERITAGE Family Study." *Journal of Applied Physiology* 88:551–559.

Boyd, R., and P. Richerson. 1988. *Culture and the Evolutionary Process.* Chicago: University of Chicago Press.

Boyd, R., and P. Richerson. 2005. *Not by Genes Alone: How Culture Transformed Human Evolution.* Chicago: University of Chicago Press.

Boyer, P. 1990. *Tradition as Truth and Communication.* Cambridge: Cambridge University Press.

Bramble, D., and D. Lieberman. 2004. "Endurance Running and the Evolution of *Homo*." *Nature* 432:345–352.

Brenowitz, E. A. 1991. "Altered Perception of Species-Specific Song by Female Birds After Lesions of a Forebrain Nucleus." *Science* 251:303–305.

Broad, K. D., J. P. Curley, and E. B. Keverne. 2006. "Mother-Infant Bonding and the Evolution of Mammalian Social Relationships." *Philosophical Transactions of the Royal Society of London B: Biological Sciences* 361:2199–2214.

Broaders, S. C., and S. Goldin-Meadow. 2009. "Truth Is at Hand: How Gesture Adds Information During Investigative Interviews." *Psychological Science* 2195:623–628.

Broca, P. 1863. "Localization des functions cerebrales. Siege du langage articule." *Bulletins de la Societe d'Anthropologie (Paris)* 4:200–203.

Brodal, A. 1981. *Neurological Anatomy.* Oxford: Oxford University Press.

Brodmann, K. 1909. *Vergleichende Lokalisationslehre der Grosshirnrinde in ithren Prinzipien dargestellt auf Grund des Zellenbaues.* Leipzig: Barth.

Brothers, L. 1990. "The Social Brain." *Concepts in Neuroscience* 1:27–51.

Brown, P., and C. D. Marsden. 1998. "What Do the Basal Ganglia Do?" *Lancet* 351:1801–1804.

Brown, S., and L. M. Parsons. 2008. "The Neuroscience of Dance." *Scientific American* 299:78–83.

Brown, S., E. Ngan, and M. Liotti. 2007. "A Larynx Area in the Human Motor Cortex." *Cerebral Cortex* 18:837–845.

Brownlee, A. 1954. "Play in Domestic Cattle in Britain: An Analysis of Its Nature." *British Veterinary Journal* 110:48–68.

Brukner, P., and K. Khan. 2006. *Clinical Sports Medicine.* Sydney: McGraw Hill.

Brunswik, E. 1943. "Organismic Achievement and Environmental Probability." *Psychological Review* 50:255–272.

Buford, K. 2010. *Native American Son: The Life and Sporting Legend of Jim Thorpe.* New York: Knopf.

Burish, M. J., H. Y. Kuesh, and S. S. H. Wang. 2004. "Brain Architecture and Social Complexity in Modern and Ancient Birds." *Brain Behavior Evolution* 63:107–124.

Burkhardt, G. 1998. "The Evolutionary Origins of Play Revisited: Lessons from Turtles." In *Animal Play: Evolutionary, Comparative, and Ecological Perspectives*, ed. M. Bekoff and J. A. Byers. Cambridge: Cambridge University Press.

Buzsaki, G. 2006. *Rhythms of the Brain*. Oxford: Oxford University Press.

Byers, J. 1998. "The Biology of Human Play." *Child Development* 69:599–600.

Byers, J., and C. Walker. 1995. "Refining the Motor Training Hypothesis for the Evolution of Play." *American Naturalist* 146:25–40.

Byrne, R. W. 1995. *The Thinking Ape: Evolutionary Origins of Intelligence*. Oxford: Oxford University Press.

Byrne, R. W., and L. A. Bates. 2007. "Sociality, Evolution, and Cognition." *Current Biology* 17:R714–R723.

Byrne, R. W., and N. Corp. 2004. "Neocortex Size Predicts Deception Rate in Primates." *Proceedings of the Royal Society of London A* 271:1693–1699.

Cacioppo, J. T., P. S. Visser, and C. L. Pickett. 2006. *Social Neuroscience*. Cambridge, Mass.: MIT Press.

Caillois, R. 1939. *Man and the Sacred*. Glencoe: University of Illinois Press, 2001.

Caillois, R. 1961. *Man, Play, and Games*. Champaign: University of Illinois Press.

Cajal, S. R. 1906. "The Structure and Connexions of Neurons." In *Nobel Lectures: Physiology or Medicine, 1901–1921*, 220–253. New York: Elsevier, 1967.

Calvin, W. 1982. "Did Throwing Stones Shape Hominid Brain Evolution?" *Ethology and Sociobiology* 3:115–124.

Calvin, W. 2001. *The Throwing Madonna: Essays on the Brain*. Lincoln, Neb.: iUniverse.

Cameron, O. G. 2002. *Visceral Sensory Neuroscience*. Oxford: Oxford University Press.

Campo, D., et al. 2013. "Whole-Genome Sequencing of Two North American Drosophila Melanogaster Populations Reveals Genetic Differentiation and Positive Selection." *Molecular Ecology* 22:5084–5097.

Camus, A. 1955. *The Myth of Sisyphus*. New York: Random House.

Cannon, W. B. 1916. *Bodily Changes in Pain, Hunger, Fear, and Rage*. New York: Appleton.

Cannon, W. B. 1927. "The James-Lange Theory of Emotions. A Critical Examination and an Alternative Theory." *American Journal of Psychology* 39:106–124.

Cannon, W. B. 1932. *The Wisdom of the Body*. New York: Norton, 1963.

Carey, S. 1985. *Conceptual Change in Childhood*. Cambridge, Mass.: MIT Press.

Carey, S. 2009. *The Origin of Concepts*. Oxford: Oxford University Press.

Carlin, J. 2008. *Playing the Enemy: Nelson Mandela and the Game That Made a Nation*. New York: Penguin.

Carter, C. S. 2007. "Sex Differences in Oxytocin and Vasopressin: Implications for Autism Spectrum Disorders?" *Behavioural Brain Research* 176:170–186.

Carter, S. C. 2014. "Oxytocin Pathways and the Evolution of Human Behavior." *Annual Review of Psychology* 65:10.1–10.23.

Carter, S. C., I. I. Lederhendler, and B. Kirkpatrick, eds. 1999. *The Integrative Neurobiology of Affiliation*. Cambridge, Mass.: MIT Press.

Chaddock, L., et al. 2010. "A Neuroimaging Investigation of the Association Between Aerobic Fitness, Hippocampal Volume, and Memory Performance in Preadolescent Children." *Brain Research* 1358:172–183.

Chaddock, L., et al. 2011. "A Review of the Relation of Aerobic Fitness and Physical Activity to Brain Structure and Function in Children." *Journal of the International Neuropsychological Society* 17:1–11.

Chalmers, N. R. 1980. "The Ontogeny of Play in Feral Olive Baboons (*Papio anubis*)." *Animal Behaviour* 28:570–585.

Chang, C. L., and S. Y. Hsu. 2004. "Ancient Evolution of Stress-Regulating Peptides in Vertebrates." *Peptides* 25:1681–1688.

Changeux, J. P. 2004. *The Physiology of Truth*. Cambridge, Mass.: Harvard University Press.

Changeux, J. P., and J. Chavaillon. 1995. *Origins of the Human Brain*. Oxford: Clarendon.

Charuvastra, A., and M. Cloitre. 2008. "Social Bonds and Posttraumatic Stress Disorder." *Annual Review of Psychology* 59:301–328.

Cheney, D. L., and R. M. Seyfarth. 1990. *How Monkeys See the World*. Chicago: University of Chicago Press.

Cheney, D. L., and R. M. Seyfarth. 2007. *Baboon Metaphysics*. University of Chicago Press.

Choleris, E., et al. 2007. "Microparticle-Based Delivery of Oxytocin Receptor Antisense DNA in the Medial Amygdala Blocks Social Recognition in Female Mice." *Proceedings of the National Academy of Sciences of the United States* 104:4670–4675.

Chomsky, N. 1965. *Aspects of the Theory of Syntax*. Cambridge, Mass.: MIT Press.

Clark, A. 1998. *Being There: Bringing Brain, Body, and World Together*. Cambridge, Mass.: MIT Press.

Clark, A. 2013. "Whatever Next? Predictive Brains, Situated Agents, and the Future of Cognitive Science." *Behavioral and Brain Sciences* 36:1–70.

Coan, J., H. Schaefer, and R. Davidson. 2006. "Lending a Hand: Social Regulation of the Neural Response to Threat." *Psychological Science* 17:1032–1039.

Coffman, K.., R. Dum, and P. Strick. 2011. "Cerebellar Vermis Is a Target of Projections from the Motor Areas in the Cerebral Cortex." *Proceedings of the National Academy of Sciences* Early Edition: 1–6.

Cohen, J. D., S. M. McClure, and A. J. Yu. 2007. "Should I Stay or Should I Go? How the Brain Manages the Tradeoff Between Exploitation and Exploration." *Philosophical Transactions of the Royal Society B* 362:933–942.

Conlon, J. M., and D. Larhammar. 2005. "The Evolution of Neuroendocrine Peptides." *General and Comparative Endocrinology* 142, nos. 1–2: 53–59.

Connor, R. C. 2007. "Dolphin Social Intelligence: Complex Alliance Relationships in Bottlenose Dolphins and a Consideration of Selective Environments for Extreme Brain Size Evolution in Mammals." *Philosophical Transactions of the Royal Society B* 362:587–602.

Cook, C. J. 2002. "Glucocorticoid Feedback Increases the Sensitivity of the Limbic System to Stress." *Physiology and Behavior* 75:455–464.

Coppens, Y. 1994. "East Side Story: The Origin of Human Kind." *Scientific American* 270:88–95.

Corballis, M. C. 2002. *From Hand to Mouth*. Princeton, N.J.: Princeton University Press.

Corballis, M. C. 2004. "The Origins of Modernity: Was Autonomous Speech the Critical Factor?" *Psychological Review* 111:543–552.

Corballis, M. C., and S. E. G. Lea. 1999. *The Descent of Mind*. Oxford: Oxford University Press.

Cosmides, L., and J. Tooby. 1992. "Cognitive Adaptations for Social Exchange." In *The Adapted Mind*, ed. J. Barkow et al. New York: Oxford University Press.

Costine, B., Oberlander, S., Davis, M., Penatti, C., Porter, D., Leaton, R., Henderson, L. 2010. "Anabolic Androgenic Steroid Exposure Alters Corticotropin Releasing Factor Expression and Anxiety-Like Behaviors in the Female Mouse. *Psychoneuroendocrinology* 35: 1473–1485.

Coubertin, P. 1896. "The Olympic Games of 1896." In *The Official Report of the Games of the First Olympiad: The Olympic Games B.C. 776–A.D. 1896*, ed. Timoleon J. Philemon, ed.

Craig, W. 1918. "Appetites and Aversions as Constituents of Instinct." *Biological Bulletin* 34:91–107.

Crespi, E. J., and R. J. Denver. 2005. "Ancient Origins of Human Developmental Plasticity." *American Journal of Human Biology* 17:44–54.

Crews, D. 1998. "The Evolutionary Antecedents of Love." *Psychoneuroendocrinology* 23:751–764.

Crews, D. 2005. "Evolution of Neuroendocrine Mechanisms That Regulate Sexual Behavior." *Trends in Endocrinology and Metabolism* 16:354–361.

Crews, D. 2008. "Epigenetics and Its Implications for Behavioral Endocrinology." *Frontiers in Neuroendocrinology* 29:344–357.

Crews, D. 2011. "Epigenetic Modifications of Brain and Behavior: Theory and Practice." *Hormones and Behavior* 59:393–398.

Culin, S. 1907. *Games of North American Indians*. Lincoln: University of Nebraska Press, 1992.

Cullinan, W. E., D. R. Ziegler, and J. P. Herman. 2008. "Functional Role of Local GABAergic Influences on the HPA Axis." *Brain Structure and Function* 213:63–72.

Cummins, D. D., and C. Allen. 1998. *The Evolution of Mind*. Oxford: Oxford University Press.

Curley, J. P., and E. B. Keverne. 2005. "Genes, Brains, and Mammalian Social Bonds." *Trends in Ecology and Evolution* 20:561–567.

Curnoe, D., and A. Thorne. 2003. "Number of Ancestral Human Species: A Molecular Perspective." *Homo* 53:201–224.

Cuvier, G. 1817. *Le regne animal*. New York: G. and C. and H. Carvill.

Dallman, M. F. 2003. "Stress by Any Other Name?" *Hormones and Behavior* 43:18–20.

Dallman, M. F., and S. Bhatnagar. 2000. *Chronic Stress and Energy Balance: Role of the Hypothalmo-Pituitary-Adrenal Axis*. New York: Oxford University Press.

Damasio, A. R. 1996. "The Somatic Marker Hypothesis and the Possible Functions of the Prefrontal Cortex." *Philosophical Transactions of the Royal Society of London* 354:1413–1420.

Darwin, C. 1859. *The Origin of Species*. New York: Mentor, 1958.

Darwin, C. 1868. *The Variation of Animals and Plants Under Domestication*. London: John Murray.

Darwin, C. 1871. *Descent of Man*. New York: Rand McNally.

Darwin, C. 1872. *The Expression of the Emotions in Man and Animals*. Oxford: Oxford University Press, 1998.

Davidson, R. J., and M. Rickman, M. 1999. "Behavioral Inhibition and the Emotional Circuitry of the Brain: Stability and Plasticity During the Early Childhood Years." In *Extreme Fear, Shyness, and Social Phobia*, ed. L. A. Schmidt and J. Schulkin. New York: Oxford University Press.

Davidson, R. J., K. M. Putnam, and C. L. Larson. 2000. "Dysfunction in the Neural Circuitry of Emotion Regulations—A Possible Prelude to Violence." *Science* 289:591–594.

Davidson, R. J., and B. McEwen. 2012. "Social Influences on Neuroplasticity: Stress and Interventions to Promote Well-Being." *Nature Neuroscience* 15:689–695.

Davis, M., D. L. Walker, and Y. Lee. 1997. "Amygdala and Bed Nucleus of the Stria Terminalis: Differential Roles in Fear and Anxiety Measured with the Acoustic Startle Reflex. *Philosophical Transactions of the Royal Society of London B: Biological Sciences* 352:1675–1687.

Davis, S. 2015. "How Stephen Curry Became the Best Shooter in the NBA." Business Insider. http://www.businessinsider.com/how-stephen-curry-became-best-shooter -in-the-nba-2015-6.

Dayan, E., and L. Cohen. 2011. "Neuroplasticity Subserving Motor Skill Learning." *Neuron* 72:443–454.

de Tocqueville, A. 1848. *De la democratie en Amerique*. Trans. by H. Mansfield and D. Winthrop, *Democracy in America* Chicago: University of Chicago Press, 2000.

Dearing, J. 2007. *The Untold Story of William G. Morgan: Inventor of Volleyball*. Livermore, Calif.: Wingspan.

DeBlock, A., and S. Dewitte. 2009. "Darwinism and the Cultural Evolution of Sports." *Perspectives in Biology and Medicine* 52:1–16.

Decety, J. 1996. "Do Imagined and Executed Actions Share the Same Neural Substrate?" *Cognitive Brain Research* 3:87–93.

Decety, J., D. Perani, and M. Jeannerod. 1994. "Mapping Motor Representations with Positron Emission Tomography." *Nature* 371:600–602.

Defelipe, J., and E. G. Jones. 1988. *Cajal on the Cerebral Cortex: An Annotated Translation of the Complete Writings*. Oxford: Oxford University Press.

Dehaene, S. 1997. *The Number Sense*. Oxford: Oxford University Press.

Dehaene, S., et al. 2006. "Core Knowledge of Geometry in an Amazonian Indigene Group." *Science* 311:381–384.

Delson, E., and K. Harvati. 2006. "Return of the Last Neanderthal." *Nature* 443:262–263.

Denton, D. 1982. *The Hunger for Salt*. Berlin: Springer-Verlag.

Desimone, R. 1996. "Neural Mechanisms for Visual Memory and Their Role in Attention." *Proceedings of the National Academy of Sciences* 93:13494–13499.

Dewey, J. 1895. "The Theory of Emotion." *Psychological Review* 2, no. 1: 13–32.

Dewey, J. 1896. "The Reflex Arc Concept in Psychology." *Psychological Review* 3:357–370.

Dewey, J. 1908. *Theory of Moral Life*. New York: Holt, Rinehart and Winston, 1960.

Dewey, J. 1910. *The Influence of Darwin on Philosophy*. Bloomington: Indiana University Press, 1965.

Dewey, J. 1925. *Experience and Nature*. LaSalle, Ill.: Open Court, 1989.

Dewey, J. 1929. *The Quest for Certainty*. New York: Capricorn, 1960.

Dewey, J. 1934a. *Art as Experience*. New York: Capricorn, 1958.

Dewey, J. 1934b. A Common Faith. New Haven: Yale University Press.

Diamond, A. 2001. "A Model System for Studying the Role of Dopamine in the Prefrontal Cortex During Early Development of Humans: Early and Continuously Treated Phenylketonuria." In *Handbook of Developmental Cognitive Neuroscience*, ed. C. A. Nelson and M. Luciana, 433–472. Cambridge, Mass.: MIT Press.

Diamond, J. 1997. *Guns, Germs, and Steel*. New York: Norton.

Diderot, D. 1755. *The Encyclopedia*. In *Rameau's Nephew and Other Works*. New York: Library of Liberal Arts, 1964.

Dinkel, D., Dhabhar, F., Sapolsky, R. 2004. "Neurotoxic effects of polymorphonuclear granulocytes on hippocampal primary cultures." *Proceedings* of the National Academy of Sciences USA 101: 331–336.

Dobzhansky, T. C. 1962. *Mankind Evolving*. New Haven, Conn.: Yale University Press.

Dolan, R. 2007. "The Human Amygdala and Orbital Prefrontal Cortex in Behavioral Regulation." *Philosophical Transactions of the Royal Society* 362:787–789.

Dolinoy, D., J. Weidman, and R. Jirtle. 2007. "Epigenetic Gene Regulation: Linking Early Developmental Environment to Adult Disease." *Reproductive Toxicology* 23:297–307.

Donald, M. 1991. *Origins of the Modern Mind*. Cambridge, Mass.: Harvard University Press.

Donald, M. 2001. *A Mind So Rare: The Evolution of Human Consciousness*. New York: Norton.

Donald, M. 2004. "Hominid Enculturation and Cognitive Evolution." In *The Development of the Mediated Mind*, ed. J. M. Luraciello et al. Mahwah, N.J.: Erlbaum.

Donaldson, Z. R., and L. J. Young. 2008. "Oxytocin, Vasopressin, and the Neurogenetics of Sociality." *Science* 322:900–904.

Donnelly, P. 2011. "From War Without Weapons to Sport for Development and Peace: The Janus-Face of Sport." *SAIS Review of International Affairs* 31:65–76.

Doral, M., et al. 2012. *Sports Injuries*. Heidelberg: Springer.

Dum, R., and P. Strick. 2013. "Transneuronal Tracing with Neurotropic Viruses Reveals Network Macroarchitecture." *Current Opinion in Neurobiology* 23:245–249.

Dunbar, R. 1992. "Neocortex Size as a Constraint on Group Size in Primates." *Journal of Human Evolution* 22:469–493.

Dunbar, R. 2010. *How Many Friends Does One Person Need?* Cambridge, Mass.: Harvard University Press.

Dunbar, R., and S. Shultz. 2007a. "Evolution in the Social Brain." *Science* 317:1344–1347.

Dunbar, R., and S. Shultz. 2007b. "Understanding Primate Brain Evolution." *Proceedings of the Royal Society B: Biological Sciences* 362:649–658.

Durkheim, E. 1974. *Sociology and Philosophy.* Trans. D. F. Pocock. New York: Free Press.

Eberhart, J. A., E. B. Keverne, and R. E. Meller. 1980. "Social Influences on Plasma Testosterone Levels in Male Talapoin Monkeys." *Hormones and Behavior* 14:247–266.

Ehlert, T., P. Simon, and D. Moser. 2013. "Epigenetics in Sports." *Sports Medicine* 43:93–110.

Eichenbaum, J., and N. J. Cohen. 2001. *From Conditioning to Conscious Recollection.* Oxford: Oxford University Press.

Ekman, P. 1972. Universals and Cultural Differences in Facial Expressions of Emotion." In *Nebraska Symposium on Motivation, 1971,* ed. J. Cole. Lincoln: University of Nebraska Press.

Eldridge, N. 1999. *The Pattern of Evolution.* New York: W. H. Freeman.

Eldridge, N. 1985. *Unfinished Synthesis.* Oxford: Oxford University Press.

Elliot, E., et al. 2010. "Resilience to Social Stress Coincides with Functional DNA Methylation of the *Crf* Gene in Adult Mice." *Nature Neuroscience* 13:1351–1353.

Ellison, T. T. 2005. "Evolutionary Perspectives on the Fetal Origins Hypothesis." *American Journal of Human Biology* 17:113–118.

Ellison, T. T., and P. B. Gray. 2009. *Endocrinology of Social Relationships.* Cambridge, Mass.: Harvard University Press.

Elster, J. 2000. *Ulysses Unbound.* Cambridge: Cambridge University Press.

Emery, N. J. 2000. "The Eyes Have It: The Neuroethology, Function, and Evolution of Social Gaze." *Neuroscience and Biobehavioral Reviews* 24:581–604.

Erickson, K. R., et al. 2009. "Aerobic Fitness Is Associated with Hippocampal Volume in Elderly Humans." *Hippocampus* 19:1030–1039.

Erickson, K. R., et al. 2011. "Exercise Training Increases Size of Hippocampus and Improves Memory." *Proceedings of the National Academy of Sciences* 108:3017–3022.

Ericsson, K., R. Krampe, and C. Tesch-Romer. 1993. "The Role of Deliberate Practice in the Acquisition of Expert Performance." *Psychological Review* 100:363–406.

Erlandson, J. 2001. "The Archaeology of Aquatic Adaptations: Paradigms for a New Millennium." *Journal of Archaeological Research* 9:287–350.

Evans, R. M. 1988. "The Steroid and Thyroid Hormone Superfamily." *Science* 240:375–402.

Everitt, B. J., and T. W. Robbins. 2005. "Neural Systems of Reinforcement for Drug Addiction: From Actions to Habits to Compulsions." *Nature Neuroscience* 11:1481–1489.

Eynon, N., et al. 2011. "Genes and Elite Athletes: A Roadmap for Future Research." *Journal of Physiology* 13:3063–3070.

Fadiga, L., L. Craighero, and A. D. Ausilio. 2009. "Broca's Area in Language, Action, and Music." *The Neurosciences and Music III–Disorders and Plasticity* 1168:448–458.

Fagen, R. 1974. "Selective and Evolutionary Aspects of Animal Play." *American Naturalist* 108:850–858.

Fagen, R. 1981. *Animal Play Behavior.* New York: Oxford University Press.

Fagen, R., and J. Fagen. 2004. "Juvenile Survival and Benefits of Play Behavior in Brown Bears, *Ursus arctos.*" *Evolutionary Ecology Research* 6:89–102.

Falk, D. 1983. "Cereberal Corticices of East African Early Hominids." *Science* 221:1072–1074.

Falk, D. 2000. "Hominid Brain Evolution and the Origins of Music." In *The Origins of Music*, ed. N. L. Wallin et al. Cambridge, Mass.: MIT Press.

Farah, M. J. 1984. "The Neurobiological Basis of Visual Imagery: A Componential Analysis." *Cognition* 18:245–272.

Faubert, J., and L. Sidebottom. 2012. "Perceptual-Cognitive Training of Athletes." *Journal of Clinical Sport Psychology* 6:85–102.

Fentress, J. C. 1984. "The Development of Coordination." *Journal of Motor Behavior* 16:99–134.

Ferguson, J., et al. 2001. "Oxytocin in the Medial Amygdala Is Essential for Social Recognition in the Mouse." *Journal of Neuroscience* 21: 8278–8285.

Finger, S. 1994. *Origins of Neuroscience.* Oxford: Oxford University Press.

Finlayson, C. 2004. *Neanderthals and Modern Humans.* Cambridge, Mass.: Cambridge University Press.

Finley, M., and H. Pleket. 2005. *The Olympic Games: The First Thousand Years.* Mineola, N.Y.: Dover.

Fiorillo, C. D., P. N. Tobler, and W. Schultz. 2003. "Discrete Coding of Reward Probability and Uncertainty by Dopamine Neurons." *Science* 299:1898–1902.

Fisher, R. A. 1930. *The Genetic Theory of Natural Selection.* Oxford: Clarendon.

Fiske, A. P. 1991. *Structures of Social Life.* New York: Free Press.

Fitzsimons, J. T. 1998. "Angiotensin, Thirst, and Sodium Appetite." *Physiological Reviews* 78:583–686.

Fliessbach, K., et al. 2007. "Social Comparison Affects Reward-Related Brain Activity in the Human Ventral Striatum." *Science* 23:1305–1308.

Foley, R. 1995. *Humans Before Humanity.* Oxford: Blackwell.

Foley, R. 1996. "An Evolutionary and Chronological Framework for Human Social Behaviour." *Proceedings of the British Academy* 88:95–117.

Foley, R. 2001. "In the Shadow of the Modern Synthesis?" *Evolutionary Anthropology* 10:5–14.

Foley, R. 2006. "The Emergence of Culture in the Context of Hominin Evolutionary Patterns." In *Evolution and Culture*, ed. S. C. Levinson and P. Jaisson. Cambridge, Mass.: MIT Press.

Ford, P., et al. "The Role of Deliberate Practice and Play in Career Progression in Sport: The Early Engagement Hypothesis." *High Ability Studies* 20:65–75.

Formenti, F., L. P. Ardigó, and A. E. Minetti. 2005. "Human Locomotion on Snow: Determinants of Economy and Speed of Skiing Across the Ages." *Proceedings of the Royal Society, Biological Sciences* 272:1561–1569.

Foster, R. G., and L. Kreitzman. 2004. *Rhythms of Life*. New Haven, Conn.: Yale University Press.

Fredrickson, B. L. 2004. "The Broaden and Build Theory of Positive Emotions." *Philosophical Transactions of the Royal Society* 359:1367–1377.

Frey, S. H. 2007. "What Puts the How in Where? Tool Use and the Divided Visual Streams Hypothesis." *Cortex* 43:368–375.

Frith, C. D. 2007. "The Social Brain?" *Proceedings of the Royal Society B: Biological Sciences* 362:671–678.

Frith, C. D., and D. Wolpert. 2003. *The Neuroscience of Social Interaction*. Oxford: Oxford University Press.

Fulton, J. F. 1949. *Functional Localization in the Frontal Lobes and Cerebellum*. Oxford: Oxford University Press.

Fuster, J. M. 2003. *Cortex and Mind*. Oxford: Oxford University Press.

Galef, B., and K. Laland. 2005. "Social Learning in Animals: Empirical Studies and Theoretical Models." *BioScience* 55:489–499.

Gallagher, M., and F. C. Holland. 1994. "The Amygdala Complex: Multiple Roles in Associative Learning and Emotion." *Proceedings of the National Academy of Sciences* 91: 11771–11776.

Gallagher, S. 2005. *How the Body Shapes the Mind*. Oxford: Oxford University Press.

Gallese, V., and A. Goldman. 1998. "Mirror Neurons and the Simulation Theory of Mind-Reading." *Trends in Cognitive Science* 2:493–501.

Gallistel, C. R. 1980. *The Organization of Action: A New Synthesis*. Hillsdale, N.J.: Lawrence Erlbaum.

Gallistel, C. R. 1990. *The Organization of Learning*. Cambridge, Mass.: MIT Press.

Gallistel, C. R., R. M. Gelman, and S. Cordes. 2005. "The Cultural and Evolutionary History of the Real Numbers." In *Culture and Evolution*, ed. S. Levinson, and P. Jaisson. Cambridge, Mass.: MIT Press.

Galton, F. 1883. *Inquiries Into Human Faculty and Its Development*. London: Macmillan.

Gardiner, N. 2002. *Athletics in the Ancient World*. Mineola, N.Y.: Dover.

Gardner, A., et al. 2013. "A Systematic Review of Concussion in Rugby League." *British Journal of Sports Medicine*.

Gazzaniga, M. S. 1985. *The Social Brain*. New York: Basic Books.

Gazzaniga, M. S. 1995. *The New Cognitive Neurosciences*. Cambridge, Mass.: MIT Press, 2000.

Gazzaniga, M. S. 2005. *The Ethical Brain*. Washington: Dana.

Geschwind, N. 1974. *Selected Papers on Language and the Brain*. Boston: Reidel.

Geschwind, N., and A. M. Galaburda. 1984. *Cerebral Dominance*. Cambridge, Mass.: Harvard University Press.

Gibbs, R. A., and D. L. Nelson. 2003. "Human Genetics: Primate Shadow Play." *Science* 28:1331–1333.

Gibson, J. J. 1979. *The Ecological Approach to Visual Perception.* Boston: Houghton Mifflin.

Gibson, K. R., and T. Ingold, eds. 1993. *Tools, Language, and Cognition in Human Evolution.* Cambridge: Cambridge University Press.

Gigerenzer, G. 2000. *Adaptive Thinking, Rationality in the Real World.* New York: Oxford University Press.

Gigerenzer, G. 2007. *Gut Feelings.* New York: Viking.

Gimpl, G., and F. Fahrenholz. 2001. "The Oxytocin Receptor System: Structure, Function, and Regulation." *Physiological Reviews* 81:629–683.

Glimcher, P. W. 2003. *Decisions, Uncertainty, and the Brain.* Cambridge, Mass.: MIT Press.

Gluckman, P. D., M. A. Hanson, and H. G. Spencer. 2005. "Predictive Adaptive Responses and Human Evolution." *Trends in Ecology and Evolution* 20:527–533.

Gold, J., and M. Gold. 2007. "Access for All: The Rise of the Paralympic Games." *Journal of the Royal Society for the Promotion of Health* 127:133–141.

Goldblatt, D. 2008. *The Ball Is Round: A Global History of Soccer.* New York: Riverhead.

Goldin-Meadow, S. 1999. "The Role of Gesture in Communication and Thinking." *Trends in Cognitive Sciences* 3:419–429.

Goldman-Rakic, P. S. 1996. "The Prefrontal Landscape: Implications of Functional Architecture for Understanding Human Mentation and the Central Executive." *Philosophical Transactions of the Royal Society of London* 351:1445–1453.

Goldsmith, R. 1940. *The Material Basis of Evolution.* New Haven, Conn.: Yale University Press, 1982.

Goldstein, D. G., and G. Gigerenzer. 2002. "Models of Ecological Rationality: The Recognition Heuristic." *Psychological Review* 109:75–90.

Goldstein, D. S. 2000. *The Autonomic Nervous System in Health and Disease.* New York: Marcel Dekker.

Goldstein, J. H. 1998. *What We Watch.* Oxford: Oxford University Press.

Goodale, M. 2014. "How (and Why) the Visual Control of Action Differs from Visual Perception." *Proceedings of the Royal Society B: Biological Sciences* 281.

Goodale, M., and K. Humphrey. 1998. "The Objects of Action and Perception." *Cognition* 67:181–207.

Goodale, M.A., and A. D. Milner. 1992. "Separate Visual Pathways for Perception and Action." *Trends in Neuroscience* 15:20–25.

Goodman, L. I., et al. "Molecular Evolution of Aerobic Energy Metabolism in Primates." *Molecular Phylogenetics and Evolution* 18:26–36.

Goodman, N. 1955. *Fact, Fiction, and Forecast.* New York: Bobbs-Merrill, 1978.

Goodson, J. L. 2005. "The Vertebrate Social Behavioral Network: Evolutionary Themes and Variations." *Hormones and Behavior* 48:11–22.

Gopnik, A., and A. Meltzoff. 1993. "Imitation, Cultural Meaning, and the Origins of Theory of Mind." *Commentary in Behavioral and Brain Sciences* 16:521–522.

Gould, E., et al. 1996. "Neurogenesis in the Dentate Gyrus of the Adult Tree Shrew Is Regulated by Psychosocial Stress and NMDA Receptor Activation." *Journal of Neuroscience* 17:2492–2498.

Gould, E., et al. 1999. "Learning Enhances Adult Neurogenesis in the Hippocampal Formation." *Nature Neuroscience* 2:260–265.

Gould, E., et al. 2001. "Adult Generated Hippocampal and Neocortical Neurons in Macaques Have a Transient Existence." *Proceedings of the National Academy of Sciences* 98:101910–101916.

Gould, J., and C. Gould. 1999. *The Animal Mind.* New York: W. H. Freeman.

Gould, S. J. 1977. *Ontogeny and Phylogeny.* Cambridge, Mass.: Harvard University Press.

Gould, S. J. 2002. *The Structure of Evolutionary Theory.* Cambridge, Mass.: Harvard University Press.

Gould, S. J. 2007. *Punctuated Equilibrium.* Cambridge, Mass.: Cambridge University Press.

Gould, S. J., and N. Eldridge. 1977. "Punctuated Equilibria: The Tempo and Mode of Evolution Reconsidered." *Paleobiology* 3:115–151.

Goy, R. W., and B. S. McEwen. 1980. *Sexual Differentiation of the Brain.* Cambridge, Mass.: MIT Press.

Grassi-Oliveria, R., and L. Stein. 2008. "Childhood Maltreatment Associated with PTSD and Emotional Distress in Low-Income Adults: The Burden of Neglect." *Child Abuse & Neglect* 32:1089–1094.

Graybiel, A. M. 1998. "The Basal Ganglia and Chunking of Action Repertoires." *Neurobiology of Learning and Memory* 70:119–136.

Greene, J.D. 2014. "The cognitive neuroscience of moral judgment and decision-making," In *The Cognitive Neurosciences V, M.S. Gazzaniga,* Ed. MIT Press.

Greene, J.D. (2014). Beyond Point-And-Shoot Morality: Why Cognitive (Neuro) Science Matters for Ethics. Ethics 124: 695–726.

Greene, J. D., and J. Haidt. 2002. "How (and Where) Does Moral Judgment Work?" *Trends in Cognitive Science* 6:517–523.

Gregory, R. L. 1973. *Eye and Brain: The Psychology of Seeing.* New York: McGraw Hill.

Griffin, D. R. 1958. *Listening in the Dark.* New Haven, Conn.: Yale University Press.

Gross, C. G. 1998. *Brain, Vision, Memory.* Cambridge, Mass.: MIT Press.

Grossmann, K. E., and K. Grossmann. 2003. "Universality of Human Social Attachment as an Adaptive Process." In *Attachment and Bonding,* ed. S. C. Carter. Cambridge, Mass.: MIT Press.

Gumbrecht, H. 2006. *In Praise of Athletic Beauty.* Cambridge, Mass.: Harvard University Press.

Guthrie, W. K. C. 1955. *The Greeks and Their Gods.* Boston: Beacon.

Guttmann, A. 1978. *From Ritual to Record.* New York: Columbia University Press.

Guttmann, A. 1984. *The Games Must Go On: Avery Brundage and the Olympic Movement.* New York: Columbia University Press.

Guttmann, A. 1986. *Sports Spectator.* New York: Columbia University Press.

Guttmann, A. 1988. *A Whole New Ballgame*. Chapel Hill: University of North Carolina Press.

Guttmann, A. 1990. *Games and Empires*. New York: Columbia University Press.

Guttmann, A. 1991. *Women's Sports: A History*. New York: Columbia University Press.

Guttmann, A. 1992. *The Olympics: History of the Modern Games*. Champaign: University of Illinois Press.

Guttmann, A. 1996. *The Erotic in Sports*. New York: Columbia University Press.

Guttmann, A. 2004. *Sports: The First Five Millennia*. Amherst: University of Massachusetts Press.

Guttmann, A. 2011. *Sports and American Art from Benjamin West to Andy Warhol*. Amherst: University of Massachusetts Press.

Guttmann, A., and L. Thompson. 2001. *Japanese Sports: A History*. Honolulu: University of Hawai'i Press.

Hacking, I. 1964. *Logic of Statistical Inference*. Cambridge: Cambridge University Press.

Hacking, I. 1975. *The Emergence of Probability*. Cambridge: Cambridge University Press.

Hacking, I. 1999. *The Taming of Chance*. Cambridge: Cambridge University Press.

Haeckel, E. 1900. *The Riddle of the Universe*. Buffalo, N.Y.: Prometheus, 1992.

Haidt, J. 2007. "The New Synthesis in Moral Psychology." *Science* 316:998–1002.

Hanson, N. R. 1958. *Patterns of Discovery*. Cambridge, Mass.: Cambridge University Press, 1972.

Hanson, N. R. 1971. *Observation and Explanation*. New York: Harper.

Harony-Nicolas, H., et al. 2014. "Brain Region-Specific Methylation in the Promoter of the Murine Oxytocin Receptor Gene Is Involved in Its Expression Regulation." *Psychoneuroendocrinology* 39:121–131.

Harvey, P., Martin, R., and Clutton-Brock, T. 1987. "Life Histories in Comparative Perspective." In *Primate Societies*, ed. B. Smuts et al., 181–196. Chicago: University of Chicago Press.

Harwell, D. 2015. "Why Hardly Anyone Sponsored the Most-Watched Soccer Match In U.S. History." *Washington Post* (July 6, 2015).

Hauk, O., I. Johnsrude, and F. Pulvermuller. 2004. "Somatotopic Representation of Action Words in Human Motor and Premotor Cortex." *Neuron* 41:301–307.

Hebb, D. O. 1949. *The Organization of Behavior*. New York: Wiley.

Heelan, P. A. 1983. *Space Perception and the Philosophy of Science*. Berkeley: University of California Press.

Heelan, P. A., and J. Schulkin. 1998. "Hermeneutical Philosophy and Pragmatism: A Philosophy of the Science." *Synthese* 115:269–302.

Heinrich, B., and R. Smolker. 1998. "Play in Common Ravens (*Corvus corax*)." In *Animal Play: Evolutionary, Comparative, and Ecological Perspectives*, ed. M. Bekoff and J. A. Byers, 27–44. Cambridge: Cambridge University Press.

Heinrichs, M., and G. Domes. 2008. "Neuropeptides and Social Behaviour: Effects of Oxytocin and Vasopressin in Humans." In *Progress in Brain Research*, ed. I. D. Neumann and R. Landgraf, 170:337–350. New York: Elsevier Science.

Helmholtz, H. 1873. *Popular Lectures in Scientific Subjects.* Trans. E. Atkinson. London: Longmans Green.

Herbert, J. 1993. "Peptides in the Limbic System: Neurochemical Codes for Coordinated Adaptive Responses to Behavioral and Physiological Demand." *Progress in Neurobiology* 41:723–791.

Herbert, J. 2015. *Testosterone.* Oxford: Oxford University Press.

Herbert, J., and J. Schulkin. 2002. "Neurochemical Coding of Adaptive Responses in the Limbic System." In *Hormones, Brain, and Behavior,* ed. D. Pfaff. New York: Elsevier.

Herman, E., et al. 2007. "Humans Have Evolved Specialized Skills of Social Cognition." *Science* 317, no. 5843: 1360–1366.

Herman, J. P., et al. 2003. "Central Mechanisms of Stress Integration." *Frontiers in Neuroendocrinology* 24:151–180.

Herrick, C. J. 1905. "The Central Gustatory Pathway in the Brain of Body Fishes." *Journal of Comparative Neurology* 15:375–486.

Hill, M., et al. 2010. "Endogenous Cannabinoid Signaling Is Essential for Stress Adaption." *Proceedings of the National Academy of Sciences* 107:9406–9411.

Hillard, K., P. Hooper, and M. Gurven. 2009. "The Evolutionary and Ecological Roots of Human Organization." *Philosophical Transactions of the Royal Society B* 364:3289–3299.

Hirst, K. 2014. "The Archaeological Study of Shell Middens: What Are Shell Middens?" http://archaeology.about.com/od/boneandivory/a/shellmidden.htm.

Hobbes, T. 1651. *Leviathan.* New York: Cambridge University Press, 1991.

Hodos, W., and A. B. Butler. 1997. "Evolution of Sensory Pathways in Vertebrates." *Brain, Behavior, and Evolution* 50:189–199.

Hofer, M. A., and R. M. Sullivan. 2001. "Toward a Neurobiology of Attachment." In *Developmental Cognitive Neuroscience,* ed. C. N. Nelson. Cambridge, Mass.: MIT Press.

Holland, L. Z., and S. Short. 2008. "Gene Duplication, Co-option, and Recruitment During the Origin of the Vertebrate Brain from the Invertebrate Chordate Brain." *Brain Behavior and Evolution* 72:91–105.

Holliday, R. 2002. "Epigenetics Comes of Age in the Twenty-First Century." *Journal of Genetics* 81:1–4.

Holliday, R., and T. Ho. 1998. "Evidence for Gene Silencing by Endogenous DNA Methylation." *Proceedings of the National Academy of Sciences USA* 95:8727–8732.

Hollinger, J. 2002. *Pro Basketball Prospectus: 2002 Edition.* Lincoln, Neb.: Potomac.

Holloway, R. 1980. "Exploring the Dorsal Surface of Hominoid Brain Endocasts by Stereoplotter and Discriminant Analysis. *Philosophical Transactions of the Royal Society of London B: Biological Sciences* 292:155–166.

Holst, E. von, and U. von St. Paul. 1963. "On the Functional Organization of Drives." *Animal Behaviour* 11:1–20.

Holst, E. von. 1973. "Relative Coordination as a Phenomenon and as a Method of Analysis of Central Nervous Functions." In *The Behavioral Physiology of Animals and Man: Selected Papers of Eric von Holst.* Coral Gables, Fla.: University of Miami Press.

Hootman, J., R. Dick, and J. Agel. 2007. "Epidemiology of Collegiate Injuries for 15 Sports: Summary and Recommendations for Injury Prevention Initiatives." *Journal of Athletic Training* 42:311–319.

Howells, W. W. 1963. *Back of History*. New York: Anchor.

Howells, W. W. 1976. "Explaining Modern Man." *Journal of Human Evolution* 5:477–495.

Hubel, D. H., and T. N. Wiesel. 2005. *Brain and Visual Perception: The Story of a Twenty-Five-Year Collaboration*. Oxford: Oxford University Press.

Humphrey, N. 1976. "The Social Function of Intellect." In *Growing Points in Ethology*, ed. P. P. G. Bateson and R. A. Hinde, 307–317. Cambridge: Cambridge University Press.

Humphrey, N. 1996. *Leaps of Faith*. New York: Basic Books.

Humphrey, N. 2007. "The Society of Selves." *Philosophical Transactions of the Royal Society of London B* 362:745–754.

Huxley, T. 1863. *Evidence as to Man's Place in Nature*. England: Williams and Norgate.

Hyodo, K., et al. 2016. "The Association Between Aerobic Fitness and Cognitive Function in Older Men Mediated by Frontal Lateralization." *NeuroImage* 125:291–300.

Ikemoto, S., and J. Panksepp. 1999. "The Role of Nucleus Accumbens Dopamine in Motivated Behavior: A Unifying Interpretation with Special Reference to Reward Seeking." *Brain Research Reviews* 31:6–41.

Inman, M. 2006. "Learning New Movements Depends on the Statistics of Your Prior Actions." *PLOS Biology* 4:e354.

Insel, T. R. 2010. "The Challenge of Translation in Social Neuroscience: A Review of Oxytocin, Vasopressin, and Affiliative Behavior." *Neuron* 65:768–779.

Insel, T. R., and R. D. Fernald. 2004. "How the Brain Processes Social Information." *Annual Review of Neuroscience* 27:697–722.

Isaac, G. L., R. E. F. Leakey, and A. K. Behrensmeyer. 1971. "Archaeological Traces of Early Hominid Activities, East of Lake Rudolf, Kenya." *Science* 173:1129–1134.

Isler, K., and C. Van Schaik. 2006. "Metabolic Costs of Brain Size Evolution." *Biology Letters* 2:557–560.

Iwaniuk, A., J. Nelson, and S. Pellis. 2001. "Do Big-Brained Animals Play More? Comparative Analysis of Play and Relative Brain Size in Mammals." *Journal of Comparative Psychology* 115:29–41.

Jablonka, E., and M. J. Lamb. 1995. *Epigenetic Inheritance and Evolution*. Oxford: Oxford University Press.

Jack, A., J. Connelly, and J. Morris. 2012. "DNA Methylation of the Oxytocin Receptor Gene Predicts Neural Response to Ambiguous Social Stimuli." *Frontiers in Human Neuroscience* 6:280.

Jackson, J. H. 1884. "Evolution and Dissolution of the Nervous System." In *Selected Writings of John Hughlings Jackson*. London: Staples, 1958.

Jackson, P. L., and J. Decety. 2004. "Motor Cognition: A New Paradigm to Self and Other Interactions." *Current Opinion in Neurobiology* 14:259–263.

Jacobowitz, D. M. 1988. "Multifactorial Control of Pituitary Hormone Secretion: The 'Wheels' of the Brain." *Synapse* 2:86–92.

Jahanson, D. C., and M. Edey. 1981. *Lucy: The Beginnings of Humankind*. New York: Simon and Schuster.

James, W. 1887. "Some Human Instincts." *Popular Science Monthly* 31:160–176.

James, W. 1890. *The Principles of Psychology*. 2 vols. New York: Dover, 1952.

Janata, P., and S. T. Grafton. 2003. "Swinging in the Brain: Shared Neural Substrates for Behaviors Related to Sequencing and Music." *Nature Neuroscience* 6:682–687.

Jancke, L. 2009. "The Plastic Human Brain." *Restorative Neurology and Neuroscience* 27:521–538.

Jarvis, E. D., and F. Nottebohm. 1997. "Motor-Driven Gene Expression." *Proceedings of the National Academy of Sciences of the United States: Neurobiology* 94:4097–4102.

Jaspers, K. 1913. *General Psychopathology*. Trans. J. Hoenig and M. W. Hamilton. Baltimore, Md.: Johns Hopkins University Press, 1997.

Jeannerod, M. 1997. *The Cognitive Neuroscience of Action*. Oxford: Blackwell.

Jeannerod, M. 1999. "To Act or Not to Act: Perspectives on the Representation of Action." *Quarterly Journal of Experimental Psychology* 52:1–29.

Jerison, H. 1979. "The Evolution of Diversity in Brain Size." In *Development and Evolution of Brain Size*, ed. M. E. Hahnet, 29–57. New York: Academic Press.

Jevning, R., A. F. Wilson, and E. F. VanderLaan. 1978. "Plasma Prolactin and Growth Hormone During Meditation." *Psychosomatic Medicine* 40:329–333.

Johanson, D. C., and M. Edey. 1981. *Lucy: The Beginnings of Humankind*. New York: Simon and Schuster.

Johnson, M. 2014. *Morality for Humans*. Chicago: University of Chicago Press.

Johnson, M. 1987. *The Body in the Mind*. Chicago: University of Chicago Press.

Johnson-Frey, S. H. 2003. "What's So Special About Tool Use?" *Neuron* 39:201–204.

Johnson-Frey, S. H., et al. 2003. "Actions or Hand-Object Interactions? Human Inferior Frontal Cortex and Action Observation." *Neuron* 39:1053–1058.

Johnston, J. B. 1923. "Further Contributions to the Study of the Evolution of the Forebrain." *Journal of Comparative Neurology* 56:337–381.

Jolly, A. 1999. *Lucy's Legacy*. Cambridge, Mass.: Harvard University Press.

Kaas, J. H. 2007. *Evolution of Nervous Systems*. Vol. 3: *Mammals*. Oxford: Elsevier.

Kaas, J. H. 2013. "The Evolution of Neocortex in Primates." *Progress in Brain Research* 195:91–102.

Kaas, J. H. 2007. "The Evolution of the Complex Sensory and Motor Systems of the Human Brain." *Brain Research Bulletin* 75:384–390.

Kagan, J. 1984. *The Nature of the Child*. New York: Basic Books.

Kagan, J. 2002. *Surprise, Uncertainty, and Mental Structure*. Cambridge, Mass.: Harvard University Press.

Kagan, J., J. S. Resnick, and N. Snidman. 1988. "Biological Bases of Childhood Shyness." *Science* 240:167–171.

Kahneman, D., P. Slovic, and A. Tversky, eds. 1982. *Judgment Under Uncertainty: Heuristics and Biases*. New York: Cambridge University Press.

Kakei, S., D. S. Hoffman, and P. L. Strick. 1999. "Muscle and Movement Representation in the Primary Motor Cortex." *Science* 285:2136–2139.

Kakei, S., D. S. Hoffman, and P. L. Strick. 2001. "Direction of Action Is Represented in the Ventral Premotor Cortex." *Nature Neuroscience* 4:1020–1025.

Kalin, N. H., S. E. Shelton, and R. J. Davidson. 2000. "Cerebrospinal Fluid Cortico-tropin-Releasing Hormone Levels Are Elevated in Monkeys with Patterns of Brain Activity Associated with Fearful Temperament." *Biological Psychiatry* 47:579–585.

Kandel, E. R., and L. R. Squire. 2000. "Neuroscience: Breaking Down Scientific Barriers to the Study of Brain and Mind." *Science* 290:1113–11120.

Kant, I. 1787. *Critique of Pure Reason*. Trans. L. W. Beck. New York: St. Martin's Press, 1965.

Kant, I. 1792. *Critique of Judgment*. New York: Haffner, 1951.

Kant, I. 1788. *Critique of Practical Reason*. Trans. L. W. Beck. New York: Bobbs-Merrill, 1956.

Kantak, S., et al. 2010. "Neural Substrates of Motor Memory Consolidation Depend on Practice Structure." *Nature Neuroscience* 13:923–925.

Kanwal, J. S., and R. D. Prasada Rao. 2002. "Oxytocin Within Auditory Nuclei: A Neuromodulatory Function in Sensory Processing?" *Neuroendocrinology* 13:2193–2197.

Kaplan, H. S., and A. J. Robson. 2002. "The Emergence of Humans: The Coevolution of Intelligence and Longevity with Intergenerational Transfers." *Proceedings of the National Academy of Sciences* 99:10221–10226.

Kaplan, H., P. Hooper, and M. Gurven. 2009. "The Evolutionary and Ecological Roots of Human Social Organization." *Proceedings of the Royal Society B: Biological Sciences* 364:3289–3299.

Kappers, C. U. A., G. C. Huber, and E. C. Crosby. 1967. *The Comparative Anatomy of the Nervous System of Vertebrates, Including Man*. New York: Hafner.

Karatsoereos, I., and B. McEwen. 2011. "Psychobiological Allostasis: Resistance, Resilience, and Vulnerability." *Trends in Cognitive Sciences* 15:576–584.

Kavushansky, A., and M. Leshem. 2004. "Role of Oxytocin and Vasopressin in the Transitions of Weaning in the Rat." *Developmental Psychobiology* 45:231–238.

Keil, F. C. 1979. *Semantic and Conceptual Development: An Ontological Perspective*. Cambridge, Mass.: Harvard University Press.

Keil, F. C. 1989. *Concepts, Kinds, and Cognitive development*. Cambridge: MIT Press.

Kelley, A. E. 1999. "Neural Integrative Activities of Nucleus Accumbens Subregions in Relation to Learning and Motivation." *Psychobiology* 27:198–213.

Kelley, D. B. 2002. "Hormonal Regulation of Motor Output in Amphibians." In *Hormones, Brain, and Behavior*, ed. D. W. Pfaff et al. New York: Academic Press.

Kempermann, G. 2006. *Adult Neurogenesis*. Oxford: Oxford University Press.

Kendrick, K. M. 2000. "Oxytocin, Motherhood, and Bonding." *Experimental Physiology* 85:111S–124S.

Kersey, R., et al. 2012. "National Athletic Trainers' Association Position Statement: Anabolic-Androgenic Steroids." *Journal of Athletic Training* 47:567–588.

Keverne, E. B., and J. P. Curley. 2008. "Epigenetics, Brain Evolution, and Behavior." *Neuroendocrinology* 29:398–412.

Kingdom, J. 2003. *Lowly Origins*. Princeton, N.J.: Princeton University Press.

Kinzler, K., E. Dupoux, and E. S. Spelke. 2007. "The Native Language of Social Cognition." *Proceedings of the National Academy of Sciences of the United States* 104:12577–12580.

Kirschner, S., and M. Tomasello. 2009. "Joint Drumming: Social Context Facilitates Synchronization in Preschool Children." *Journal of Experimental Child Psychology* 102:299–314.

Kirschner, S., and M. Tomasello. 2010. "Joint Music Making Promotes Prosocial Behavior in Four-Year-Old Children." *Evolution and Human Behavior* 31:354–364.

Klein, R. G. 1989. *The Human Career*. Chicago: University of Chicago Press.

Klein, R. G. 2008. "Out of Africa and the Evolution of Human Behavior." *Evolutionary Anthropology* 17:267–281.

Knoblich, G., and N. Sebanz. 2006. "The Social Nature of Perception." *Current Directions in Psychological Science* 15:99–104.

Knowlton, B., J. Mangels, and L. Squire. 1996. "A Neostriatal Habit Learning System in Humans." *Science* 273:1399–1402.

Knutson, B. J., et al. 2008. "Nucleus Accumbens Activation Mediates the Influence of Reward Cues on Financial Risk Taking." *Brain Imaging* 19:509–513.

Knutson, B., J. Burgdorf, and J. Panksepp. 1998. "Anticipation of Play Elicits High Frequency Ultrasonic Vocalizations in Young Rats." *Journal of Comparative Psychology* 112:65–73.

Koehl, M., et al. 2008. "Exercise-Induced Promotion of Hippocampal Cell Proliferation Requires B-endorphin." *FASEB Journal* 22:2253–2262.

Konorski, J. 1967. *Integrative Activity of the Brain*. Chicago: University of Chicago Press.

Koob, G. F., and M. LeMoal. 2008. "Addiction and the Brain Antireward System." *Annual Review of Psychology* 59:29–53.

Koob, G. F., and M. LeMoal. 2005. *Neurobiology of Addiction*. New York: Elsevier.

Koob, G. F., et al. 1994. "Corticotropin-Releasing Factor, Stress, and Behavior." *Seminars in Neuroscience* 6:221–229.

Kornblith, H. 1993. *Inductive Inference and Its Natural Ground*. Cambridge, Mass.: MIT Press.

Kosslyn, S. M. 1986. *Image and Mind*. Cambridge, Mass.: Harvard University Press.

Krettek, J. E., and J. L. Price. 1977. "Projections from the Amygdaloid Complex to the Cerebral Cortex and Thalamus in the Rat and Cat." *Journal of Comparative Neurology* 172:687–722.

Kringelbach, M. L., and K. C. Berridge. 2010. *Pleasures of the Brain*. Oxford: Oxford University Press.

Kripke, S. 1980. *Naming and Necessity*. Cambridge, Mass.: Harvard University Press.

Kumsta, R., et al. 2013. "Epigenetic Regulation of the Oxytocin Receptor Gene: Implications for Behavioral Neuroscience." *Frontiers in Neuroscience* 7:1–6.

Lakoff, G., and M. Johnson. 1999. *Philosophy in the Flesh: The Embodied Mind and Its Challenge to Western Thought*. New York: Basic Books.

Lakoff, G., and R. E. Numez. 2000. *Where Mathematics Comes From*. New York: Basic Books.

Lamarck, J. B. 1809. *Zoological Philosophy*. Trans. H. Elliot. Chicago: University of Chicago Press, 1984.

Landau, B., and J. E. Hoffman. 2005. "Parallels Between Spatial Cognition and Spatial Language: Evidence from Williams Syndrome." *Journal of Memory and Language* 53:163–185.

Lange, C., and E. Irle. 2004. "Enlarged Amygdala Volume and Reduced Hippocampal Volume in Young Women with Major Depression." *Psychological Medicine* 34:1059–1064.

Langer, S. K. 1953. *Feeling and Form*. New York: Scribner.

LaRocca, T., D. Seals, and G. Pierce. 2010. "Leukocyte Telomere Length Is Preserved with Aging in Endurance Exercise-Trained Adults and Related to Maximal Aerobic Capacity." *Mechanisms of Aging and Development* 131:165–167.

Larramendi, L. M. H. 1969. "Analysis of Synaptogenesis in the Cerebellum of the Mouse." In *Neurobiology of Cerebellar Evolution and Development*, ed. R. R. Llinas, 803–843. Chicago: AMA.

Lashley, K. S. 1938. "An Experimental Analysis of Instinctive Behavior." *Psychological Review* 45:445–471.

Lashley, K. S. 1951. "The Problem of Serial Order in Behavior." In *Cerebral Mechanisms in Behavior*, ed. L. A. Jeffres, 110–133. New York: Wiley.

Laughlin, S. 2001. "Energy as a Constraint on the Coding and Processing of Sensory Information." *Current Opinion in Neurobiology* 11:475–480.

Lazarus R. S. 1984. "On the Primacy of Cognition." *American Psychologist* 39:124–129.

Lazarus, R. S. 1991. *Emotion and Adaptation*. New York: Oxford University Press.

Leakey, L. S. B. 1958. "Recent Discoveries at Olduvai Gorge, Tanganyika." *Nature* 181:1099–1103.

Leakey, L. S. B, and M. D. Leakey. 1964. "Recent Discoveries of Fossil Hominids in Tanganyika, at Olduvai and Near Lake Natron." *Nature* 202:5–7.

Leakey, L. S. B., and R. Lewin. 1977. *Origins: The Emergence and Evolution of Our Species and Its Possible Future*. New York: Penguin.

LeDoux, J. E. 1995. "Emotion: Clues from the Brain." *Annual Review of Psychology* 46:209–235.

LeDoux, J. E. 2000. "Emotion Circuits in the Brain." *Annual Review of Neuroscience* 23:155–184.

LeDoux, J. E. 2012. "Rethinking the Emotional Brain." *Neuron* 73:653–676.

LeDoux, J.E. 2016. *Anxious*. New York: Penguin Random House.

Leigh, S. R. 2004. "Brain Growth, Life History, and Cognition in Primate and Human Evolution." *American Journal of Human Evolution* 62:139–164.

Leslie, A., R. Gelman, and C. R. Gallistel. 2008. "The Generative Basis of Natural Number Concepts." *Trends in Cognitive Science* 12:213–218.

Levi, I. 1967. *Gambling with Truth*. Cambridge, Mass.: MIT Press.

Levinson, S. 2006. "Cognition at the Heart of Human Interaction." *Discourse Studies* 8:85–93.

Levinson, S., and P. Jaisson, eds. 2006. *Evolution and Culture*. Cambridge, Mass.: MIT Press.

Lewis, K. P., and R. A. Barton. 2006. "Amygdala Size and Hypothalamus Predict Social Play Frequency in Nonhuman Primates." *Journal of Comparative Psychology* 120:31–37.

Lewis, M. 2004. *Moneyball: The Art of Winning an Unfair Game*. New York: Norton.

Lieberman, D. 2012. "Human Evolution: Those Feet in Ancient Times." *Nature* 483:550–551.

Lieberman, D. E. 2007. "Homing in on Early *Homo*." *Nature* 499:291–292.

Lieberman, D. E. 2011. *The Evolution of the Human Head*. Cambridge, Mass.: Harvard University Press.

Lieberman, D. E. 2013. *The Story of the Human Body: Evolution, Health, and Disease*. New York: Random House.

Lieberman, M. D. 2006. "Intuition: A Social Cognitive Neuroscience Approach." *Psychological Bulletin* 126:109–117.

Lieberman, P. 1984. *The Biology and Evolution of Language*. Cambridge, Mass.: Harvard University Press.

Lieberman, P. 2002. *Human Language and Our Reptilian Brain*. Cambridge, Mass.: Harvard University Press.

Lieberman, P. 2009. "FOXP2 and Human Cognition." *Cell* 137:800–802.

Lieberman, P., and R. McCarthy. 2007. "Tracking the Evolution of Language and Speech." *Expedition* 49:15–20.

Lim, M. M., I. F. Bielsky, and L. J. Young. 2005. "Neuropeptides and the Social Brain." *International Journal of Developmental Neuroscience* 23:235–243.

Lincoln, A. 1861. First Inaugural Address as President of the United States. Washington DC.

Lindenfors, P. 2005. "Neocortex Evolution in Primates: The Social Brain Is for Females." *Biology Letters* 1:407–410.

Lippi, G., U. Longo, and N. Maffulli. 2010. "Genetics and Sports." *British Medical Bulletin* 93:27–47.

Liu, Y., and Z. X. Wange. 2003. "Nucleus Accumbens Oxytocin and Dopamine Interact to Regulate Pair Bond Formation in Female Prairie Voles." *Neuroscience* 121:537–544.

Loewenstein, G. F. 1994. "The Psychology of Curiosity." *Psychological Bulletin* 116:75–98.

Loewenstein, G. F. 2006. "The Pleasures and Pains of Information." *Science* 312:704–706.

Lorenz, K. 1981. *The Foundations of Ethology*. New York: Springer.

Lovejoy, D. A., and R. J. Balment. 1999. "Evolution and Physiology of the Corticoropin-Releasing Factor (CRF) Family and Neuropeptides in Vertebrates." *General and Comparative Endocinology* 115:1–22.

Lovejoy, D. A., and S. Jahan. 2006. "Phylogeny of Corticotrophin-Releasing Factor Family of Peptides in the Metazoan." *General and Comparative Endocrinology* 146:1–8.

Lovett, C. 1997. *Olympic Marathon: A Centennial History of the Games and Most Storied Race*. Westport, Conn.: Praeger.

Macaloon, J. 2007. *This Great Symbol: Pierre de Coubertin and the Origins of the Modern Olympic Games*. London: Routledge.

Maclean, P. D. 1990. *The Triune Brain in Evolution*. New York: Plenum.

Mahler, M. 2000. *The Psychological Birth of the Human Infant: Symbiosis and Individuation*. New York: Basic Books.

Mahler, S., and K. Berridge. 2009. "Which Cue to 'Want?' Central Amygdala Opioid Activation Enhances and Focuses Incentive Salience on a Prepotent Reward Cue." *Journal of Neuroscience* 29:6500–6513.

Makino, S., P. W. Gold, and J. Schulkin. 1994. "Corticosterone Effects on Corticotropin-Releasing Hormone mRNA in the Central Nucleus of the Amygdala and the Parvocellular Region of the Paraventricular Nucleus of the Hypothalamus." *Brain Research* 640:105–112, 141–149.

Mandell, R. 1971. *The Nazi Olympics*. Champaign: University of Illinois Press.

Mandell, R. 1984. *Sport: A Cultural History*. New York: Columbia University Press, 1999.

Mangan, J., ed. 2004. *Militarism, Sport, Europe: War Without Weapons*. London: Frank Cass.

Manzon, L. A. 2002. "The Role of Prolactin in Fish Osmoregulation: A Review." *General and Comparative Endocrinology* 125:291–310.

Marino, L. 2007. "The Evolution of the Brain and Cognition in Cetaceans." In *Evolutionary Cognitive Neuroscience*, ed. S. M. Platek et al. Cambridge, Mass.: MIT Press.

Marler, C. A., S. K. Boyd, and W. Wilczynski. 1999. "Forebrain Arginine Vasotocin Correlates of Alternative Mating Strategies in Cricket Frogs." *Hormones and Behavior* 36:53–61.

Marler, P., et al. 1988. "The Role of Sex Steroids in the Acquisition and Production of Birdsong." *Nature* 336:770–772.

Martin, A. 2007. "The Representation of Object Concepts in the Brain." *Annual Review of Psychology* 58:25–45.

Martin, A. 2015. "GRAPES—Grounding Representations in Action, Perception, and Emotion Systems: How Object Properties and Categories Are Represented in the Human Brain." *Psychometric Bulletin & Review*. In press.

Matsuzaka, Y., N. Picard, and P. Strick. 2007. "Skill Representation in the Primary Motor Cortex After Long-Term Practice." *Journal of Neurophysiology* 97:1819–1832.

Mayr, E. 1942. *Systemics and the Origin of Species*. New York: Columbia University Press, 1982.

Mayr, E. 1963. *Animal Species and Evolution*. Cambridge, Mass.: Harvard University Press.

McCrea, M., et al. 2003. "Concussion in Collegiate Football Players: The NCAA Concussion Study." *Journal of the American Medical Association* 290:2556–2563.

McCrory, P. 2009. "Sport Concussion Assessment Tool 2." *Scandinavian Journal of Medicine and Science in Sports* 19:452.

McEwen, B. S. 1998. "Protective and Damaging Effects of Stress Mediators." *New England Journal of Medicine* 338:171–179.

McEwen, B. S. 2007. "Physiology and Neurobiology of Stress and Adaptation: Central Role of the Brain." *Physiological Reviews* 87:873–904.

McHenry, H. M. 1994. "Tempo and Mode in Human Evolution." *Proceedings of the National Academy of Sciences* 91:6780–6786.

McHenry, H. M. 2009. "Human Evolution." In *Evolution: The First Four Billion Years*, ed. Michael Ruse and Joseph Travis, 256–280. Cambridge, Mass.: Harvard University Press.

McKee, A., et al. 2009. "Chronic Traumatic Encephalopathy in Athletes: Progressive Tauopathy Following Repetitive Head Injury." *Journal of Neuropathology and Experimental Neurology* 68:709–735.

Mead, G. H. 1934. *Mind, Self, and Society: From the Standpoint of a Social Behaviorist.* Chicago: University of Chicago Press.

Meaney, M. J. 2001. "Maternal Care, Gene Expression, and the Transmission of Individual Differences in Stress Reactivity Across Generations." *Annual Review of Neuroscience* 24:1161–1192.

Mellars, P. 2006a. "Going East: New Genetic and Archaeological Perspectives on the Modern Human Colonization of Eurasia." *Science* 313:796–800.

Mellars, P. 2006b. "Why Did Modern Human Populations Disperse from Africa 60,000 Years Ago?" *Proceedings of the National Academy of Sciences* 103:9381–9386.

Meltzoff, A. N., and M. K. Moore. 1977. "Imitation of Facial and Manual Gestures by Human Neonates." *Science* 198:75–78.

Merleau-Ponty, M. 1942. *The Structure of Behavior.* Boston: Beacon, 1967.

Merleau-Ponty, M. 1962. *The Phenomenology of Perception.* New York: Routledge and Kegan Paul.

Meyer,-Lindenberg, A. 2008. "Impact of Prosocial Neuropeptides on Human Brain Function." *Progress in Brain Research* 170:463–470.

Miller, N. E. 1957. "Experiments on Motivation. Studies Combining Psychological, Physiological, and Pharmacological Techniques." *Science* 126:1271–1278.

Milner, D., and M. A. Goodale. 1995. *The Visual Brain in Action.* Oxford: Oxford University Press.

Ming, G. L., and H. Song. 2005. "Adult Neurogenesis in the Mammalian Central Nervous System." *Annual Review of Neuroscience* 28:223–250.

Missitzi, J., et al. 2013. "Heritability of Motor Control and Motor Learning." *Physiological Reports* 1:1–10.

Mithen, S. 1996. *The Prehistory of the Mind.* London: Thames and Hudson.

Mogenson, G. J. 1987. "Limbic-Motor Integration." In *Progress in Psychobiology and Physiological Psychology*, ed. A. N. Epstein and J. Sprague. New York: Academic Press.

Moll, H., and M. Tomasello. 2007. "Cooperation and Human Cognition: The Vygotskian Intelligence Hypothesis." *Proceedings of the Royal Society B: Biological Sciences* 362:639–648.

Moll, J., and J. Schulkin. 2009. "Social Attachment and Aversion: On the Humble Origins of Human Morality." *Neuroscience and Biobehavioral Reviews* 33: 456-465.

Moore, F. L. 1992. "Evolutionary Precedents for Behavioral Actions of Oxytocin and Vasopressin." *Annals of the New York Academy of Sciences* 652:156–165.

Moran, T. 2000. "Cholecystokinin and Satiety: Current Perspectives." *Nutrition* 16:858–865.

Morgan, C. A., et al. 2004. "Relationships Among Plasma Dehydroepiandrosterone Sulfate and Cortisol Levels, Symptoms of Dissociation, and Objective Performance in Humans Exposed to Acute Stress." *Archives of General Psychiatry* 61:819–825.

Morris, J. S., et al. 1996. "A Differential Neural Response in the Human Amygdala to Fearful and Happy Facial Expressions." *Nature* 383:812–815.

Mountcastle, V. B. 1998. *Perceptual Systems.* Cambridge, Mass.: Harvard University Press.

Mrosovsky, N. 1990. *Rheostasis: The Physiology of Change.* New York: Oxford University Press.

Muller, S., B. Abernethy, and D. Farrow. 2006. "How Do World-Class Cricket Batsmen Anticipate a Bowler's Intention? *Quarterly Journal of Experimental Psychology* 59:2612–2186.

Mullestin, L., et al. 2012. "Genetic Influences on Physical Activity in Young Adults: A Twin Study." *Medicine and Science in Sports and Exercise* 44, no. 7: 1293–1301.

Munck, A., P. M. Guyre, and N. J. Holbrook. 1984. "Physiological Functions of Glucocorticoids in Stress and Their Relations to Pharmacological Actions." *Endocrine Reviews* 5:25–44.

Murray, E. A., and S. P. Wise. 2004. "What, If Anything, Is the Medial Temporal Lobe, and How Can the Amygdala Be Part of It If There Is No Such Thing?" *Neurobiology of Learning and Memory* 82:178–198.

Murray, S. 2010. "Boxing Gloves of the Ancient World." *Journal of Combative Sport.*

Mustard, J. A., K. T. Beggs, and A. R. Mercer. 2005. "Molecular Biology of the Invertebrate Dopamine Receptors." *Archives of Insect Biochemistry and Physiology* 59:103–117.

Naismith, J. 1941. *Basketball: Its Origin and Development.* New York: Association Press.

Nelson, E., and J. Panksepp. 1996. "Oxytocin Mediates Acquisition of Maternally Associated Odor Preferences in Preweanling Rat Pups." *Behavioral Neuroscience* 110:583–592.

Neumann, I., and R. Landgraf. 2012. "Balance of Brain Oxytocin and Vasopressin: Implications of Anxiety, Depression, and Social Behaviors." *Trends in Neurosciences* 35:649–654.

Newkirk, P. 2009. *Spectacle: The Astonishing Life of Ota Benga.* New York: Harper Collins.

Noe, A. 2004. *Action in Perception.* Cambridge, Mass.: MIT Press.

Norgren, R. 1995. "Gustatory System." In *The Rat Nervous System,* 2nd ed., 751–771. San Diego, Calif.: Academic Press.

O'Bleness, M., et al. 2012. "Evolution of Genetic and Genomic Features Unique to the Human Lineage." *Nature Reviews Genetics* 13:853–866.

O'Connor, R. C. 2007. "Dolphin Social Intelligence: Complex Alliance Relationships in Bottlenose Dolphins and a Consideration of Selective Environments for Extreme Brain Size Evolution in Mammals." *Philosophical Transactions of the Royal Society of London B: Biological Sciences* 362:587–602.

O'Doherty, J., et al. 2004. "Dissociable Roles of Ventral and Dorsal Striatum in Instrumental Conditioning." *Science* 304:452–454.

Omalu, B., et al. 2005. "Chronic Traumatic Encephalopathy in a National Football League Player." *Neurosurgery* 57:128–134.

Owen, C., A. Howard, and D. Binder. 2009. "Hippocampus Minor, Calcar Avis, and the Huxley-Owen Debate." *Heurosurgery* 65:1098–1104.

Palombit, R. A., R. M. Seyfarth, and D. L. Cheney. 1997. "The Adaptive Value of 'Friendships' to Female Baboons: Experimental and Observational Evidence." *Animal Behavior* 54:599–614.

Panksepp, J. 1998. *Affective Neuroscience: The Foundations of Human and Animal Emotions*. New York: Oxford University Press.

Panskepp, J. 2004. "The Ontogeny of Play in Rats." *Developmental Psychology* 14:327–332.

Papez, J. W. 1937. "A Proposed Mechanism of Emotion." *Archives of Neurological Psychiatry* 38:725–743.

Parrott, W. G., and J. Schulkin. 1993. "Neuropsychology and the Cognitive Nature of Emotions." *Cognition and Emotion* 7:43–59.

Pashler, H. 1998. *The Psychology of Attention*. Cambridge, Mass.: MIT Press.

Passingham, R. E. 1993. *The Frontal Lobes and Voluntary Action*. Oxford: Oxford University Press, 1997.

Passingham, R. E. 2008. *What Is Special About the Human Brain*? Oxford: Oxford University Press.

Passmore, J., and C. Gibbes. 2007. "The State of Executive Coaching Research: What Does the Current Literature Tell Us and What's Next for Coaching Research?" *International Coaching Psychology Review* 2, no. 2: 116–128.

Payne, M. 2015. "The Artificial Turf at the Women's World Cup Was Reportedly 120 Degrees at Kick Off." *Washington Post* (June 6, 2015).

Peciña, S., K. S. Smith, and K. C. Berridge. 2006. "Hedonic Hot Spots in the Brain." *Neuroscientist* 12:500–511.

Peirce, C. S. 1868. "Questions Concerning Certain Faculties Claimed for Man." *Journal of Speculative Philosophy* 2:103–114.

Peirce, C. S. 1878. "Deduction, Induction, and Hypothesis." *Popular Science Monthly* 13:470–482.

Pellis, S. M. 1991. "How Motivationally Distinct Is Play? A Preliminary Case Study." *Animal Behaviour* 42:851–853.

Perret, D. I., and N. J. Emery. 1994. "Understanding the Intentions of Others from Visual Signals: Neurophysiological Evidence." *Cahiers de Pscyhologie Cognitive* 13:683–694.

Peterson, C., and H. Wellman. 2009. "From Fancy to Reason: Scaling Deaf and Hearing Children's Understanding of Theory of Mind and Pretence." *British Journal of Developmental Psychology* 27:297–310.

Phillips, L., et al. 2013. "Postprandial Total and HMW Adiponectin Following a High-Fat Meal in Lean, Obese, and Diabetic Men." *European Journal of Clinical Nutrition* 2, no. 67: 377–384.

Piaget, J. 1954. *The Construction of Reality in the Child*. New York: Basic Books.

Pickford, G., and E. Strecker. 1977. "The Spawning Reflex Response of the Killifish *Fundulus heteroclitus*; Isotocin Is Relatively Inactive in Comparison with Arginine Vasotocin." *General and Comparative Endocrinology* 32:132–137.

Pinker, S. 1994. *The Language Instinct*. New York: William Morrow.

Pinker, S. 2011. *The Better Angels Of Our Nature*. New York: Viking.

Pinnock, S., and J. Herbert. 2009. "Brain-Derived Neurotropic Factor and Neurogenesis in the Adult Rate Dentate Gyrus: Interactions with Corticosterone." *European Journal of Neuroscience* 27:2493–2500.

Piomelli, D. 2003. "The Molecular Logic of Endocannabinoid Signalling." *Nature Reviews Neuroscience* 4, no. 11: 873–884.

Plato, Anastaplo, G., Berns, L. 2004. Meno. Newburyport, MA: Focus Pub./R. Pullins Co.

Pontzer, H., D. Raichlen, and M. Sockol. 2009. "The Metabolic Cost of Walking in Humans, Chimpanzees, and Early Hominins." *Journal of Human Evolution* 56:43–54.

Pontzer, H., and R. Wrangham. 2004. "Climbing and the Daily Energy Cost of Locomotion in Wild Chimpanzees: Implications for Hominoid Locomotor Evolution." *Journal of Human Evolution* 46:315–333.

Power, M. L., and J. Schulkin. 2009. *Evolution of Obesity*. Baltimore, Md.: Johns Hopkins University Press.

Power, T. G. 2000. *Play and Exploration in Children and Animals*. Mahwah, N.J.: Lawrence Erlbaum Associates.

Powley, T. L. 1977. "The Ventralmedial Hypothalamic Syndrome, Satiety and Cephalic Phase." *Psychological Review* 84:89–126.

Powley, T. L. 2000. "Vagal Input to the Enteric Nervous System." *Gut* 47:iv30–iv32.

Premack, D. 1990. "The Infant's Theory of Self-Propelled Objects." *Cognition* 36:1–16.

Premack, D., and A. J. Premack. 1994. "Moral Belief: Form Versus Content." In *Mapping the Mind: Domain Specificity in Cognition and Culture*, ed. L. A. Hirschfeld et al. New York: Cambridge University Press.

Premack, D., and A. J. Premack. 1995. "Origins of Human Social Competence." In *The Cognitive Neurosciences*, ed. M. S. Gazzaniga, 205–218. Cambridge, Mass.: MIT Press.

Premack, D., and G. Woodruff. 1978. "Does the Chimpanzee Have a Theory of Mind?" *Behavioral and Brain Sciences* 1:515–526.

Prinz, J. L., and L. W. Barsalou. 2000. "Steering a Course for Embodied Representation." In *Cognitive Dynamics: Conceptual Change in Humans and Machines*, ed. E. Dietrich and A. B. Markman. Cambridge, Mass.: MIT Press.

Puterman, E., et al. 2010. "The Power of Exercise: Buffering the Effect of Chronic Stress on Telomere Length." *PLoS One* 5:e10837–e20601.

Quaglio, G., et al. 2009. "Anabolic Steroids: Dependence and Complications of Chronic Use." *Internal Emergency Medicine* 4:289–296.

Rader, B. 2009. *American Sports: From the Age of Folk Games to the Age of Televised Sports, 6/E*. Upper Saddle River, N.J: Pearson.

Raichlen, D., and A. Gordon. 2011. "Relationship Between Exercise Capacity and Brain Size in Mammals." *PLoS One* 6:e20601.

Raichlen, D., et al. 2013. "Exercise-Induced Endocannabinoid Signaling Is Modulated by Intensity." *European Journal of Applied Physiology* 113:869–875.

Raichlen, D., Armstrong, H., Lieberman, D. (2011). "Calcaneus length determines running economy: Implications for endurance running performance in modern humans and Neandertals." *Journal of Human Evolution* 60: 299–308.

Rampersad, A. 1998. *Jackie Robinson: A Biography*. New York: Ballantine.

Rampon, C., et al. 2000. "Effects of Environmental Enrichment on Gene Expression in the Brain." *Proceedings of the National Academy of Sciences* 97:12880–12884.

Rathelot, J., and P. Strick. 2009. "Subdivisions of Primary Motor Cortex Based on Cortico-Motonneuronal Cells." *Proceedings of the National Academy of Sciences* 106:918–923.

Rauch, H., G. Schonbachler, and T. Noakes. 2013. "Neural Correlates of Motor Vigour and Motor Urgency During Exercise." *Sports Medicine* 43:227–241.

Rauschecker, J. P., and M. Korte. 1993. "Auditory Compensation for Early Blindness in Cat Cerebral Cortex." *Journal of Neuroscience* 139:4538–4548.

Rauschecker, J. P., and S. K. Scott. 2009. "Maps and Streams in the Auditory Cortex: Nonhuman Primates Illuminate Human Speech Processing." *Nature Neuroscience* 12:718–724.

Rawls, J. 1971. *A Theory of Justice*. Cambridge, Mass.: Harvard University Press.

Reader, S. M., and K. N. Laland. 2002. "Social Intelligence, Innovation, and Enhanced Brain Size in Primates." *Proceedings of the National Academy of Sciences* 99:4436–4441.

Reddy, R. 2011. "A Comparative Study of Speed Among Foot Ball Players and Hockey Players of Kakatiya University." *International Journal of Health, Physical Education, and Computer Science in Sports* 4:86–87.

Reep, R. L., B. L. Finlay, and R. B. Darlington. 2007. "The Limbic System in Mammalian Brain Evolution." *Brain, Behavior, Evolution* 70:57070.

Reiner A., L. Medina, and C. L. Veenman. 1998. "Structural and Functional Evolution of the Basal Ganglia in Vertebrates." *Brain Research Reviews* 28:235–285.

Rescorla, R. A., and A. R. Wanger. 1972. "A Theory of Pavlovian Conditioning: Variations in the Effectiveness of Reinforcement Nonreinforcment." In *Classical Conditioning: Current Research and Theory*, ed. W. J. Baker and W. Prokasy. New York: Appleton-Century-Crofts.

Richerson, P. J., and R. Boyd. 2005. *Not by Genes Alone*. Chicago: University of Chicago Press.

Richter, C. P. 1943. *Total Self-Regulatory Functions in Animals and Man*. New York: Harvey Lecture Series.

Richter, C.P. 1949. "Domestication of the Norway Rat and Its Implications for the Problem of Stress." *Proceedings of the Association for Research in Nervous and Mental Disease* 29:19–47.

Richter, C. P. 1965. *Biological Clocks in Medicine and Psychiatry*. Springfield, IL: C.C. Thomas, 1979.

Riefenstahl, L. 1935. *Triumph des Willens*. Germany: Reichsparteitag-Film.

Riefenstahl, L. 1938. *Olympia*. Gemany: Olympia-Film.

Rilling, J., and T. Insel. 1998. "Evolution of the Cerebellum in Primates: Differences in Relative Volume Among Monkeys, Apes, and Humans." *Brain, Behaviour, and Evolution* 52:308–314.

Rizzolatti, G., and G. Luppino. 2001. "The Cortical Motor System." *Neuron* 31:889–901.

Rizzolatti, G., and M. A. Arbib. 1998. "Language Within Our Grasp." *Trends in Neuroscience* 21:188–194.

Roach, N., et al. 2012. "The Effect of Humeral Torsion on Rotational Range of Motion." *Journal of Anatomy* 220:293–301.

Roach, N., et al. 2013. "Elastic Energy Storage in the Shoulder and the Evolution of High-Speed Throwing in *Homo*." *Nature* 498:483–486.

Robson, S., and B. Wood. 2008. "Hominin Life History: Reconstruction and Evolution." *Journal of Anatomy* 212:394–425.

Rolls, E. T. 2000. "The Orbitofrontal Cortex and Reward." *Cerebral Cortex* 10:284–294.

Rolls, E. T., and A. Treves. 1998. *Neural Networks and Brain Function.* New York: Oxford University Press.

Rorty, R. 2000. *Philosophy and Social Hope.* New York: Penguin.

Rosenzwieg, M. R. 1984. "Experience, Memory, and the Brain." *American Psychologist* 39:365–375.

Rousseau, J. 1762. *Du contrat social ou Principes du droit politique.* Paris: Jean-Jacques Rousseau.

Rousseau, J. 1782/1980. *Reveries of a Solitary Walker.* London: Penguin Classics

Rozin, P. 1976. "The Evolution of Intelligence and Access to the Cognitive Unconscious." In *Progress in Psychobiology and Physiological Psychology,* ed. J. Sprague and A. N. Epstein. New York: Academic Press.

Rozin, P. 1998. "Evolution and Development of Brains and Cultures: Some Basic Principles and Interactions." In *Brain and Mind: Evolutionary Perspectives,* ed. M. S. Gazzaniga and J. S. Altman. Strasbourg: Human Frontiers Science Program.

Runciman, W. G., J. M. Smith, and R. I. M. Dunbar, eds. 1996. *Evolution of Social Behavior Patterns in Primates and Man.* New York: Oxford University Press.

Sabini, J., and M. Silver. 1982. *Moralities of Everyday Life.* Oxford: Oxford University Press.

Sacks, O. 2015. *On the Move: A Life.* New York: Knopf.

Salen, K., and E. Zimmerman. 2005. *The Game Design Reader: A Rules of Play Anthology.* Cambridge, Mass.: MIT Press.

Sanghani, R. 2015. "Muslim Gymnast Criticized for 'Revealing' Leotard as She Wins Double-Gold." *Telegraph* (June 18, 2015).

Saper, C. 1995. "Central Autonomic System." In *The Rat Nervous System,* ed. G. Paxinos, 107–135. New York: Academic Press.

Sapolsky, R. M. 1992. *Stress: The Aging Brain and the Mechanisms of Neuron Death.* Cambridge, Mass.: MIT Press.

Schacter, D. L., and E. Tulving. 1994. *Memory Systems.* Cambridge, Mass.: MIT Press.

Schaffer, K., and S. Smith, eds. 2000. *The Olympics at the Millennium: Power, Politics, and the Games.* New Brunswick, N.J.: Rutgers University Press.

Schonberg, T., et al. 2012. "Decreasing Ventromedial Prefrontal Cortex Activity During Sequential Risk Taking: An fMRI Investigation of the Balloon Analog Risk Task." *Frontiers in Neuroscience* 6:1–11.

Schulkin, J. 1999. *The Neuroendocrine Regulation of Behavior*. Cambridge, Mass.: Cambridge University Press.

Schulkin, J. 2000. *Roots of Social Sensibility*. Cambridge, Mass.: MIT Press.

Schulkin, J. 2003. *Rethinking Homeostasis*. Cambridge, Mass.: MIT Press.

Schulkin, J. 2004. *Bodily Sensibility: Intelligent Action*. Oxford: Oxford University Press.

Schulkin, J. 2006. *Effort: A Behavioral Neuroscience Perspective on the Will*. Mahwah, N.J.: Lawrence Erlbaum.

Schulkin, J. 2009. *Cognitive Adaptation: A Pragmatist Perspective*. Cambridge, Mass.: Cambridge University Press.

Schulkin, J. 2011. *Adaptation and Well-Being: Social Allostasis*. Cambridge: Cambridge University Press.

Schultz, W. 2002. "Getting Formal with Dopamine and Reward." *Neuron* 36:241–263.

Schultz, W. 2007. "Multiple Dopamine Functions at Different Time Courses." *Annual Review of Neuroscience* 30:259–288.

Schumacher, M., and F. Robert. 2002. "Progesterone: Synthesis, Metabolism, Mechanisms of Action, and Effects in the Nervous System." In *Hormones, Brain, and Behavior*, ed. D. Pfaff et al., 3:683–745. San Diego, Calif.: Academic Press.

Schumann, C., and D. Amaral. 2005. "Stereological Estimation of the Number of Neurons in the Human Amygdaloid Complex." *Journal of Comparative Neurology* 491:320–329.

Schutz, A. 1932. *The Phenomenology of the Social World*. Trans. G. Walsh and F. Lehnert. Chicago: Northwestern University Press, 1967.

Seyfarth, R., and D. Cheyney. 1984. "Grooming, Alliances, and Reciprocal Altruism in Vervet Monkeys." *Nature* 308:541–543.

Shadmehr, R., and J. Krakauer. 2008. "A Computational Neuroanatomy for Motor Control." *Experimental Brain Research* 185:359–381.

Sherrington, C. 1906. *The Integrative Action of the Nervous System*. New Haven, Conn.: Yale University Press.

Sherwood, N. M., and D. B. Parker. 1990. "Neuropeptide Families: An Evolutionary Perspective." *Journal of Experimental Zoology Supplement* 4:63–71.

Shors, T. J., et al. 2001. "Neurogenesis in the Adult Is Involved in the Formation of Trace Memories." *Nature* 410:372–375.

Shultz, S., and R. I. M. Dunbar. 2006. "Both Social and Ecological Factors Predict Ungulate Brain Size." *Proceedings of the Royal Society B: Biological Sciences* 273:207–215.

Shultz, S., E. Nelson, and R. I. M. Dunbar. 2012. "Hominin Cognitive Evolution: Identifying Patterns and Processes in the Fossil and Archaeological Record." *Philosophical Transactions of the Royal Society B* 367:2130–2140.

Silk, J. B. 2007. "The Adaptive Value of Sociality in Mammalian Groups." *Philosophical Transactions of the Royal Society B* 362:539–559.

Silver, N. 2012. *The Signal and the Noise: Why Most Predictions Fail—but Some Don't*. New York: Penguin.

Siviy, S. 2010. "Play and Adversity: How the Playful Mammalian Brain Withstands Threats and Anxieties." *American Journal of Play* 2:297–314.

Siviy, S., and J. Panskepp. 2011. "In Search of Neurobiological Substrates for Social Playfulness in Mammalian Brains." *Neuroscience and Biobehavioral Reviews* 35:1821–1830.

Smeets, J., et al. 2006. "Sensory Integration Does Not Lead to Sensory Calibration." *Proceedings of the National Academy of Sciences* 103:18781–18786.

Smith, A. 1759. *The Theory of Moral Sentiments*. Edinburgh: A. Millar, A. Kincaid, and J. Bell.

Smith, H. 1953. *From Fish to Philosopher*. New York: Little, Brown.

Sol, D., et al. 2008. "Brain Size Predicts the Success of Mammal Species Introduced Into Novel Environments." *American Naturalist* 172:S63–S71.

Spelke, E. 1990. "Principles of Object Perception." *Cognitive Science* 14:29–56.

Squire, L. R. 1987. *Memory and Brain*. New York: Oxford University Press.

Squire, L. R. 2004. "Memory Systems of the Brain: A Brief History and Current Perspective." *Neurobiology of Learning and Memory* 82:171–177.

Squire, L., B. Knowlton, and G. Musen. 1993. "The Structure and Organization of Memory." *Annual Review of Psychology* 44:453–495.

Stelter, R. 2007. "Coaching: A Process of Personal and Social Meaning Making." *International Coaching Psychology* 2:191–212.

Stephan, H., H. Frahm, and G. Baron. 1987. "Comparison of Brain Structure Volumes in *Insectivora* and Primates. VII. Amygdaoloid Components." *Journal fur Hirnforschung* 28:571–584.

Sterelny, K. 2007. "Social Intelligence, Human Intelligence, and Niche Construction." *Philosophical Transactions of the Royal Society of London B* 362:719–730.

Sterling, P. 2004. "Principles of Allostasis: Optimal Design, Predictive Regulation, Psychopathology, and Rational Therapeutics." In *Allostasis, Homeostasis, and the Costs of Physiological Adaptation*, ed. J. Schulkin. Cambridge, Mass.: Cambridge University Press.

Sterling, P., and J. Eyer. 1988. "Allostasis: A New Paradigm to Explain Arousal Pathology." In *Handbook of Life Stress, Cognition, and Health*, ed. S. Fisher and J. Reason. New York: John Wiley and Sons.

Sterling, P., and S. Laughlin. 2015. *Principles of Neural Design*. Cambridge: MIT Press.

Stone, A. 2002. *Heat of Creation: The Mesoamerican World and the Legacy of Linda Schele*. Tuscaloosa: University of Alabama Press.

Strand, F. L. 1999. *Neuropeptides: Regulators of Physiological Processes*. Cambridge, Mass.: MIT Press.

Stringer, C. B. 1992. "Reconstructing Recent Human Evolution." *Philosophical Transactions of the Royal Society of London B* 337:217–224.

Stringer, C. B., and P. Andrews. 1988. "Genetic and Fossil Evidence for the Origin of Modern Humans." *Science* 239:1263–1268.

Sutton-Smith, B. 1994. "Paradigms of Intervention." In *Play and Intervention*, ed. J. Hellendorn et al. Albany: State University of New York Press.

Sutton-Smith, B. 1997. *The Ambiguity of Play*. Cambridge, Mass.: Harvard University Press.

Swanson, L. W. 2000. "Cerebral Hemisphere Regulation of Motivated Behavior." *Brain Research* 886:113–164.

Swanson, L. W. 2003. *Brain Architecture*. Oxford: Oxford University Press.

Symons, D. 1978. "Play and Aggression: A Study of Rhesus Monkeys." New York: Columbia University Press.

Szymanski, S. 2006. "The Economic Evolution of Sport and Broadcasting." *Australian Economic Review* 39:428–434.

Tattersall, I. 1993. *The Human Odyssey: Four Million Years of Human Evolution*. New York: Prentice Hall.

Tattersall, I., and J. Schwartz. 2001. *Extinct Humans*. Boulder, Colo.: Westview.

Thornton, J. 2001. "Evolution of Vertebrate Steroid Receptors from an Ancestral Estrogen Receptor by Ligand Exploitation and Serial Genome Expansions." *Proceedings of the National Academy of Sciences* 98:5671–5676.

Thorpe, S. K. S., R. L. Holder, and R. H. Crompton. 2007. "Origin of Human Bipedalism as an Adaptation for Locomotion on Flexible Branches." *Science* 316:1328–1331.

Tinbergen, N. 1951. *The Study of Instinct*. Oxford: Oxford University Press, 1969.

Todes, D. 2014. *Ivan Pavlov: A Russian Life in Science*. Oxford: Oxford University Press.

Tomasello, M. 1999. *The Cultural Origins of Human Cognition*. Cambridge, Mass.: Harvard University Press.

Tomasello, M., and M. Carpenter. 2007. "Shared Intentionality." *Developmental Science* 12:121–125.

Tomasello, M., A. Kruger, and H. Ratner. 1993. "Cultural Learning." *Behavioral and Brain Sciences* 16:495–552.

Toohey, K., and A. Veal. 1999. *The Olympic Games: A Social Science Approach*. New York: CABI.

Tran, L., et al. 2014. "Epigenetic Modulation of Chronic Anxiety and Pain by Histone Deacetylation." *Molecular Psychiatry*: 1–13.

Tremblay, L., R. Hollerman, and W. Schultz. 1998. "Modifications of Reward Expectation–Related Neuronal Activity During Learning in Primate Striatum." *Journal of Neurophysiology* 80:964–977.

Tulving, E., and F. I. M. Craik. 2000. *The Oxford Handbook of Memory*. Oxford: Oxford University Press.

Ullman, M. T. 2001. "A Neurocognitive Perspective on Language: The Declarative Procedural Model." *Nature Neuroscience* 9:266–286.

Ullman, M. T. 2004. "Is Broca's Area Part of a Basal Ganglia Thalamocortical Circuit?" *Cognition* 92:231–270.

Ungerleider, L. G., and M. Mishkin. 1982. "Two Cortical Visual Systems." In *Analysis of Visual Behavior*, ed. D. Ingle et al. Cambridge, Mass.: MIT Press.

Urgesi, C., et al. 2012. "Long- and Short-Term Plastic Modeling of Action Prediction Abilities in Volleyball." *Psychological Research* 76:542–560.

Uvnäs-Moberg, K. 1998. "Oxytocin May Mediate the Benefits of Positive Social Interaction and Emotions." *Psychoneuroendocrinology* 23:819–835.

Van Essen, D. C. 2005. "Corticocortical and Thalamocortical Information Flow in the Primate Visual System." *Progress in Brain Research* 149:173–185.

Van Essen, D. C., C. H. Anderson, and D. J. Felleman. 1992. "Information Processing in the Primate Visual System: An Integrated Systems Perspective." *Science* 255:419–422.

Vanderschuren, L., R. Niesink, and J. Van Ree. 1997. "The Neurobiology of Social Play Behavior in Rats." *Neuroscience and Biobehavioral Reviews* 21:309–326.

Varela, F. J., E. Thompson, and E. Rosch. 1991. *The Embodied Mind.* Cambridge, Mass.: MIT Press.

Veenema, A., and I. D. Neumann. 2008. "Central Vasopressin and Oxytocin Release: Regulation of Complex Social Behaviors." *Progress in Brain Research* 70:261–267.

Vyas, A., Mitra, R., Rao, B., Chattarji, S. 2002. "Chronic stress induces contrasting patterns of dendritic remodeling in hippocampal and amygdaloid neurons." *Journal of Neuroscience* 22: 6810–68128.

Walker, B. 2007. *The Anatomy of Sports Injuries.* Chichester: Lotus.

Walls, M. 2012. "Kayak Games and Hunting Enskilment: An Archaeological Consideration of Sports and the Situated Learning of Technical Skills." *World Archaeology* 44:175–188.

Watson, J., and F. Crick. 1953. "Molecular Structure of Nucleic Acids. A Structure for Deoxyribose Nucleic Acid." *Annals of the New York Academy of Sciences* 758:13–14.

Weaver, I. C., et al. 2004. "Epigenetic Programming by Maternal Behavior." *Nature Neuroscience* 7:847–854.

Weibel, E. R., et al. 2004. "Allometric Scaling of Maximal Metabolic Rate in Mammals: Muscle Aerobic Capacity as Determinant Factor." *Respiratory Physiology and Neurobiology* 140:115–132.

Wernicke, C. 1874. *Der aphasische symptomencomplex: Eine psychologische studie auf anatomischer basis.* Breslau: Kohn und Weigert.

White, T. D., G. Suwa, and B. Asfaw. 1995. "*Australopithecus ramidus*: A New Species of Early Hominid from Aramis, Ethiopia." *Nature* 375:88.

Whitehead, A. N. 1919. *An Enquiry Concerning the Principles of Natural Knowledge.* New York: Dover, 1982.

Whitehead, A. N. 1929. *The Function of Reason.* Boston: Beacon, 1958.

Whitehead, A. N. 1938. *Modes of Thought.* New York: Free Press, 1967.

Wilson, J. S. 1993. *The Moral Sense.* New York: Basic Books.

Wilson, M. 2002. "Six Views of Embodied Cognition." *Psychonomic Bulletin and Review* 9:625–636.

Wingfield, J. C., and L. M. Romero. 2001. "Adrenocortical Responses to Stress and Their Modulation in Free-Living Vertebrates." In *Handbook of Physiology*, sec. 7: *The*

Endocrine System, vol. 4: *Coping with the Environment: Neural and Endocrine Systems*, ed. B. S. McEwen. Oxford: Oxford University Press.

Wingfield, J. C., et al. 1999. "Testosterone, Aggression, and Communication: Ecological Bases of Endocrine Phenomena." In *The Design of Animal Communication*, ed. M. D. Hauser and M. Konishi. Cambridge, Mass.: Cambridge University Press.

Wise, R. A., and P. P. Rompre. 1989. "Brain Dopamine and Reward." *Annual Review of Psychology* 40:191–225.

Wise, R. A. 2005. "Forebrain Substrates of Reward and Motivation." *Journal of Comparative Neurology* 493:115–121.

Wise, S. P. 1985. "The Primate Premotor Cortex: Past, Present, and Preparatory." *Annual Review of Neuroscience* 8:1–19.

Wise, S. P. 2006. "The Ventral Premotor Cortex, Corticospinal Region C, and the Origin of Primates." *Cortex* 42:521–524.

Wittgenstein, L. 1953. *Philosophical Investigations*. New York: MacMillan, 1968.

Wood, B. 1992. "Origin and Evolution of the Genus *Homo*." *Nature* 355:783–790.

Wood, R., and S. Stanton. 2012. "Testosterone and Sport: Current Perspectives." *Hormones and Behavior* 61:147–155.

Woods, S. C., R. A. Hutton, and W. Makous. 1970. "Conditioned Insulin Secretion in the Albino Rat." *Proceedings of the Society of Experimental Biology and Medicine* 133:965–968.

Wrangham, R. 1987. "The Evolution of Social Structure." In *Primate Societies*, ed. B. B. Smuts et al., 282–296. Chicago: University of Chicago Press.

Wynn, T., and F. L. Coolidge. 2008. "A Stone-Age Meeting of Minds." *American Scientist* 96:44–51.

Yamamoto, K., and P. Vernier. 2011. "The Evolution of Dopamine Systems in Chordates." *Frontiers in Neuroanatomy* 5:1–21.

Yarrow, K., P. Brown, and J. Krakauer. 2009. "Inside the Brain of an Elite Athlete: The Neural Processes That Support High Achievement in Sports." *Nature Reviews Neuroscience* 10:585–596.

Yong, E. 2013. "Origin of Domestic Dogs." *The Scientist*. http://www.thescientist. com/?articles.view/articleNo/38279/title/Origin-of-Domestic-Dogs.

Young, A. W. 1998. *Face and Mind*. Oxford: Oxford University Press.

Zak, P. J., R. Kurzban, and W. T. Matzner. 2005. "Oxytocin Is Associated with Human Trustworthiness." *Hormones and Behavior* 48:522–527.

Zimmer, C. 2005. *Smithsonian Intimate Guide to Human Origins*. Washington, D.C.: Smithsonian Books.

INDEX